Collected Works of René Guénon

The Metaphysical Principles
of the Infinitesimal Calculus

RENÉ GUÉNON

The Metaphysical Principles of the Infinitesimal Calculus

SOPHIA PERENNIS

Originally published in French as
Les Principes du Calcul Infinitésimal
© Éditions Gallimard 1946
English translation © Sophia Perennis 2001
First English Edition 2003
Second Impression 2004
Third Impression 2025

Series editor: James R. Wetmore

For information, address:
Sophia Perennis, P.O. Box 931
Philmont NY 12565

ISBN 978-0-900588-12-9 (pbk)
ISBN 978-0-900588-08-2 (cloth)

The publisher gives special thanks to
Henry D., Jennie L., & Samuel D. Fohr
for making this edition possible

Contents

Editorial Note

The past century has witnessed an erosion of earlier cultural values as well as a blurring of the distinctive characteristics of the world's traditional civilizations, giving rise to philosophic and moral relativism, multiculturalism, and dangerous fundamentalist reactions. As early as the 1920s, the French metaphysician René Guénon (1886–1951) had diagnosed these tendencies and presented what he believed to be the only possible reconciliation of the legitimate (although apparently conflicting) demands of outward religious forms, 'exoterisms', with their essential core, 'esoterism'. His works are characterized by a foundational critique of the modern world coupled with a call for intellectual reform; a renewed examination of metaphysics, the traditional sciences, and symbolism, with special reference to the ultimate unanimity of all spiritual traditions; and finally, a call to the work of spiritual realization. Despite their wide influence, translation of Guénon's works into English has so far been piecemeal. The *Sophia Perennis* edition is intended to fill the urgent need to present them in a more authoritative and systematic form. A complete list of Guénon's works, given in the order of their original publication in French, follows this note.

Guénon's early and abiding interest in mathematics, like that of Plato, Pascal, Leibniz, and many other metaphysicians of note, runs like a scarlet thread throughout his doctrinal studies. In this late text published just five years before his death, Guénon devotes an entire volume to questions regarding the nature of limits and the infinite with respect to the calculus both as a mathematical discipline and as symbolism for the initiatic path. This book therefore extends and complements the geometrical symbolism he employs in other works, especially *The Symbolism of the Cross*, *The Multiple States of the Being*, and *Symbols of Sacred Science*.

According to Guénon, the concept 'infinite number' is a contradiction in terms. Infinity is a metaphysical concept at a higher level of reality than that of quantity, where all that can be expressed is the indefinite, not the infinite. But although quantity is the only level recognized by modern science, the numbers that express it also possess qualities, their quantitative aspect being merely their outer husk. Our reliance today on a mathematics of approximation and probability only further conceals the 'qualitative mathematics' of the ancient world, which comes to us most directly through the Pythagorean-Platonic tradition.

Guénon often uses words or expressions set off in 'scare quotes'. To avoid clutter, single quotation marks have been used throughout. As for transliterations, Guénon was more concerned with phonetic fidelity than academic usage. The system adopted here reflects the views of scholars familiar both with the languages and Guénon's writings. Brackets indicate editorial insertions, or, within citations, Guénon's additions. Wherever possible, references have been updated, and English editions substituted.

This translation of *The Metaphysical Principles of the Infinitesimal Calculus* was made specifically for the Collected Works of René Guénon edition. The series editor would like to acknowledge the invaluable assistance in this project of Michael Allen, Henry Fohr, Richard Pickrell, and mathematician and traditionalist author Dr. Wolfgang Smith. Latin translations were provided by David Matz.

The Works of René Guénon

Introduction to the Study
of the Hindu Doctrines (1921)

Theosophy: History of
a Pseudo-Religion (1921)

The Spiritist Fallacy (1923)

East and West (1924)

Man and His Becoming
according to the Vedānta (1925)

The Esoterism of Dante (1925)

The Crisis of the Modern World
(1927)

The King of the World (1927)

Spiritual Authority and
Temporal Power (1929)

The Symbolism of the
Cross (1931)

The Multiple States of the Being
(1932)

The Reign of Quantity and
the Signs of the Times (1945)

Perspectives on Initiation (1946)

The Great Triad (1946)

The Metaphysical Principles of
the Infinitesimal Calculus
(1946)

Initiation and Spiritual
Realization (1952)

Insights into Christian
Esoterism (1954)

Symbols of Sacred Science
(1962)

Studies in Freemasonry
and the Compagnonnage (1964)

Studies in Hinduism (1966)

Traditional Forms and Cosmic
Cycles (1970)

Insights into Islamic Esoterism
and Taoism (1973)

Reviews (1973)

Miscellanea (1976)

Preface

Although the present study might, at least at first glance, appear to have only a rather 'specialist' character, the undertaking seemed worthwhile in order to clarify and explain more thoroughly various notions to which we have had recourse on various occasions when we have made use of mathematical symbolism, and this reason alone would suffice to justify it. However, we should add that there are still other, secondary reasons, concerning especially what one could call the 'historical' aspect of the question; the latter, indeed, is not entirely devoid of interest from our point of view inasmuch as all the discussions that have arisen on the subject of the nature and value of the infinitesimal calculus offer a striking example of that absence of principles which characterizes the profane sciences, that is, the only sciences that the moderns know and even consider possible. We have already often noted that most of these sciences, even insofar as they still correspond to some reality, represent no more than simple, debased residues of some of the ancient, traditional sciences: the lowest part of these sciences, having ceased to have contact with the principles, and having thereby lost its true, original significance, eventually underwent an independent development and came to be regarded as knowledge sufficient unto itself, although in truth it so happens that its own value as knowledge is thereby reduced to almost nothing. This is especially apparent with the physical sciences, but as we have explained elsewhere,[1] in this respect modern mathematics itself is no exception if one compares it to what was for the ancients the science of numbers and geometry; and when we speak here of the ancients one must understand by that even those of 'classical' antiquity, as the least study of Pythagorean and Platonic theories suffices to show, or at least should show were it not necessary to take into account the extraordinary

---

1. See *The Reign of Quantity and the Signs of the Times.*

1

incomprehension of those who claim to interpret them today. Were this incomprehension not so complete, how could one maintain, for example, a belief in the 'empirical' origin of the sciences in question? For in reality they appear on the contrary all the more removed from any 'empiricism' the further back one goes in time, and this is equally the case for all other branches of scientific knowledge.[2]

Mathematicians of modern times, and more particularly still those who are our contemporaries, seem to be ignorant of what number truly is; and by this we do not mean to speak solely of number in the analogical and symbolic sense as understood by the Pythagoreans and Kabbalists, which is all too obvious, but—and this might seem stranger and almost paradoxical—even of number in its simply and strictly quantitative sense. Indeed, their entire science is reduced to calculation in the narrowest sense of the word,[3] that is, to a mere collection of more or less artificial procedures, which are in short only valuable with respect to the practical applications to which they give rise. Basically this amounts to saying that they replace number with the numeral; and furthermore, this confusion of the two is today so widespread that one could easily find it at any moment, even in the expressions of everyday language.[4] Now a numeral is, strictly speaking, no more than the clothing of a number; we do not even say its body, for it is rather the geometric form that can, in certain respects, legitimately be considered to constitute the true body of a number, as the theories of the ancients on polygons and polyhedrons show when seen in the light of the symbolism of numbers; and this, moreover, is in accordance with the fact that all 'embodiment' necessarily implies a 'spatialization'. We do not mean to say, however, that numerals themselves are entirely arbitrary signs, the form of which has been determined only by the fancy of one or more individuals; there must be both numerical

---

2. See *Miscellanea*, pt. 3, chap. 1. ED.

3. The French *calcul* has the double meaning of 'calculus' and 'calculation'. ED.

4. It is the same with certain 'pseudo-esoterists', who know so little of what they wish to speak about that they likewise never fail to confuse the two in the fanciful ravings they presume to substitute for the traditional science of numbers!

and alphabetical characters—the two of which, moreover, are not distinguished in some languages[5]—and one can apply to the one as well as to the other the notion of a hieroglyphic, that is to say an ideographic or symbolic origin, and this holds for all writing without exception, however obscured this origin might be in some cases due to more or less recent distortions or alterations.

What is certain is that in their notation mathematicians employ symbols the meaning of which they no longer understand, and which are like vestiges of forgotten traditions; and what is more serious, not only do they not ask themselves what this meaning might be, it even seems that they do not want them to have any meaning at all. Indeed, they tend more and more to regard all notation as simple 'convention', by which they mean something set out in an entirely arbitrary manner, but this is a true impossibility, for one never establishes a convention without having some reason for doing so, and for doing precisely that rather than anything else; it is only to those who ignore this reason that the convention can appear as arbitrary, just as it is only to those who ignore the cause of an event that it can appear 'fortuitous'. This is indeed what occurs here, and one can see in it one of the more extreme consequences of the absence of principles, which can even cause the science—or what is so called, for at this point it no longer merits the name in any respect—to lose all plausible significance. Moreover, by the very fact of the current conception of science as exclusively quantitative, this 'conventionalism' has gradually spread from mathematics to the more recent theories of the physical sciences, which thus distance themselves further and further from the reality they intend to

5. Hebrew and Greek are two examples, and Arabic was equally so before the introduction of the use of numerals of Indian origin, which then, being more or less modified, passed from there to Europe in the Middle Ages; in this connection one can note that the word 'cipher' [French *chiffre*, 'numeral'] is itself none other than the Arabic *ṣifr*, though this word is in reality only the designation for zero. On the other hand, it is true that in Hebrew *saphar* means 'to count' or 'to number', and at the same time 'to write', whence *sepher*, 'scripture' or 'book' (in Arabic *sifr*, which designates in particular a sacred book), and *sephar*, 'numeration' or 'calculation'; from this last word also comes the designation of the *Sephiroth* of the Kabbalah, which are the principial 'numerations' assimilated to the divine attributes.

explain; we have emphasized this point sufficiently enough in another work to be able to dispense with further remarks in this regard, and all the more so since we now intend to occupy ourselves more particularly with mathematics alone. From this viewpoint we will only add that when one completely loses sight of the meaning of a notation it becomes all too easy to pass from a legitimate and valid use of it to one that is illegitimate and in fact no longer corresponds to anything, and which can sometimes even be entirely illogical. This may seem rather extraordinary when it is a question of a science like mathematics which should have particularly close ties with logic, yet it is nevertheless all too true that one can find multiple illogicalities in mathematical notions as they are commonly envisaged in our day.

One of the most remarkable examples of these illogical notions, and which we shall consider first and foremost, even though it is certainly not the only one we shall encounter in the course of our exposition, is that of the so-called mathematical or quantitative infinite, which is the source of almost all the difficulties that can be raised against the infinitesimal calculus, or, perhaps more precisely, against the infinitesimal method, for we here have something that, whatever the 'conventionalists' might think, goes beyond the range of a simple 'calculation' in the ordinary sense of the word; and this notion is the source of all difficulties without exception, save those that proceed from an erroneous or insufficient conception of the notion of the 'limit', which is indispensable if the rigor of the infinitesimal method is to be justified and made anything more than a simple method of approximation. As we shall see, moreover, there is a distinction to be made between cases in which the so-called infinite is only an absurdity pure and simple, that is, an idea contradictory in itself, such as that of an 'infinite number', and cases in which it is only employed in an improper way in the sense of indefinite; but it should not be believed because of this that the confusion of the infinite and the indefinite can itself be reduced to a mere question of words, for it rests quite truly with the ideas themselves. What is singular is that this confusion, which had it once been dispelled would have cut short so many discussions, is found in the writings of Leibniz himself, who is generally regarded as the inventor of the

4

infinitesimal calculus, although we would rather call him its 'formulator', for his method corresponds to certain realities that, as such, have an existence independent of those who conceive of them and who express them more or less perfectly; realities of the mathematical order, like all other realities, can only be discovered and not invented, while on the contrary it is indeed a question of 'invention' when, as occurs all too often in this field, one allows oneself to be swept away by the 'game' of notation into the realm of pure fantasy. But it would assuredly be quite difficult to make some mathematicians understand this difference, since they willingly imagine that the whole of their science is and must be no more than a 'fabrication of the human mind', which, if we had to believe them, would certainly reduce their science to a trifling thing indeed. Be that as it may, Leibniz was never able to explain the principles of his calculus clearly, and this shows that there was something in it that was beyond him, something that was as it were imposed upon him without his being conscious of it; had he taken this into account, he most certainly would not have engaged in any dispute over 'priority' with Newton. Besides, these sorts of disputes are always completely vain, for ideas, insofar as they are true, are not the property of anyone, despite what modern 'individualism' might have to say; it is only error that can properly be attributed to human individuals. We shall not elaborate further on this question, which could take us quite far from the object of our study, although in certain respects it would perhaps not be profitless to make it clear that the role of those who are called 'great men' is to a great extent often a role of 'reception', though they are generally the first to delude themselves as to their own 'originality'.

What concerns us more directly for the moment is this: if we must point out such deficiencies in Leibniz—deficiencies all the more serious in that they bear above all on questions of principles— what could be said of those found in other modern philosophers and mathematicians, to whom Leibniz is certainly superior in spite of everything? This superiority he owes on the one hand to the studies he made of the Scholastic doctrines of the Middle Ages, even though he did not always fully understand them, and on the other hand to certain esoteric data, principally of a Rosicrucian origin or inspiration,[6] data obviously very incomplete and even fragmentary, which he moreover sometimes applied quite poorly, as we shall presently see in some examples. It is to these two 'sources', to speak as the historians do, that one can definitively relate nearly all that is really valid in his theories, and this also allowed him to react, albeit imperfectly, against the Cartesianism which, in the double domain of philosophy and science, represented the whole ensemble of the tendencies and conceptions that are most specifically modern. This remark suffices, in short, to explain in a few words all that Leibniz was, and if one seeks to understand him, one must never lose sight of this general information, which we have for this reason deemed worthwhile to set forth at the outset; but it is time to leave these preliminary considerations in order to enter into the examination of the very questions that will allow us to determine the true significance of the infinitesimal calculus.

6. The undeniable mark of this origin is to be found in the Hermetic figure placed by Leibniz at the head of his treatise *De Arte Combinatoria*: it is a representation of the *Rota Mundi*, in which, at the center of the double cross of the elements (fire and water, air and earth) and qualities (hot and cold, dry and moist), the *quinta essentia* is symbolized by a rose with five petals (corresponding to ether considered in itself and as principle of the four other elements); naturally, this 'signature' has been passed over completely by all academic commentators.

1

Infinite
and Indefinite

Proceeding in a manner inverse to that of profane science, and in accordance with the unchanging perspective of all traditional science, we must before all else set forth the principle that will allow us almost immediately to resolve the difficulties to which the infinitesimal method has given rise, without letting ourselves be led astray by potentially interminable discussions, as indeed happens in the case of those modern philosophers and mathematicians who, by the very fact that they lack this principle, have never provided a satisfactory and definitive solution to these difficulties. This principle is the very idea of the Infinite, understood in its only true sense, which is the purely metaphysical sense, and on this subject, moreover, we have only summarily to recall what we have already expressed more completely elsewhere:[1] the Infinite is properly that which has no limits, for 'finite' is obviously synonymous with 'limited'; one cannot then correctly apply this term to anything other than that which has absolutely no limits, that is to say the universal All, which includes in itself all possibilities and consequently cannot be limited by anything in any way; the Infinite, thus understood, is metaphysically and logically necessary, for not only does it not imply any contradiction, not enclosing within itself anything negative, but it is on the contrary its negation that would be contradictory. Furthermore, there can obviously be only one Infinite, for two supposedly distinct infinites would limit and therefore inevitably exclude one another; consequently, every time the term 'infinite' is used in any sense

1. *The Multiple States of the Being*, chap. 1.

7

other than that which we have just mentioned, we can be assured *a priori* that this use is necessarily improper, for it amounts in short either to ignoring the metaphysical Infinite altogether, or to supposing another Infinite alongside it.

It is true that the Scholastics admitted what they called the *infinitum secundum quid* [the infinite in a certain respect], and that they carefully distinguished it from the *infinitum absolutum* [the absolute infinite], which alone is the metaphysical Infinite; but we can see here only an imperfection in their terminology, for although this distinction allowed them to escape the contradiction of a plurality of infinites understood in the proper sense, the double use of the word *infinitum* nonetheless certainly risked causing multiple confusions, and besides, one of the two meanings was then altogether improper, for to say that something is infinite only in a certain respect—and this is the exact significance of the expression *infinitum secundum quid*—is to say that in reality it is not infinite at all.[2] Indeed, it is not because a thing is not limited in a certain sense or in a certain respect that one can legitimately conclude that it is limited in no way at all, the latter being necessary for it to be truly infinite; not only can it be limited in other respects at the same time, but we can even say that it is of necessity so, inasmuch as it is a determined thing, which, by its very determination, does not include every possibility, and this amounts to saying that it is limited by that which lies outside of it; if, on the contrary, the universal All is infinite, this is precisely because there is nothing that lies outside of it.[3] Therefore every determination, however general one supposes it to be and however far one extends the term, necessarily excludes the true notion of the infinite;[4] a determination, whatever it might be, is always a limitation, since its essential character is to define a certain

2. It is in a rather similar sense that Spinoza later used the expression 'infinite in its kind', which naturally gives rise to the same objections.

3. One could say further that it leaves outside itself only the impossible, which, being a pure nothing, could not limit it in any way.

4. This is equally true for determinations of a universal and no longer simply general order, including even Being itself, which is the first of all determinations; but it goes without saying that this consideration does not enter into the uniquely cosmological applications we are dealing with in the present study.

domain of possibilities in relation to all the rest, and by that very fact to exclude all the rest. Thus it is truly 'nonsense' to apply the idea of the infinite to any given determination, as for example, in the instance we are considering more particularly here, to quantity or to one or another of its modes. The idea of a 'determined infinite' is too manifestly contradictory for us to dwell upon any longer, although this contradiction has most often escaped the profane thought of the moderns; and even those whom one might call 'semi-profane',[5] like Leibniz, were unable to perceive it clearly. In order to bring out the contradiction still further we could say in other fundamentally equivalent terms that it is obviously absurd to wish to define the Infinite, since a definition is in fact nothing other than the expression of a determination, and the words themselves show clearly enough that what is subject to definition can only be finite or limited. To seek to place the Infinite within a formula, or, if one prefer, to clothe it in any form whatsoever is, consciously or unconsciously, to attempt to fit the universal All into one of its minutest parts, and this is assuredly the most manifest of impossibilities.

What we have just said suffices to establish, without leaving room for the slightest doubt and without necessitating any other considerations that there cannot be a mathematical or quantitative infinite, and that this expression does not even have any meaning, because quantity is itself a determination. Number, space, and time, to which some people wish to apply the notion of this so-called infinite, are determined conditions, and as such can only be finite; they are but certain possibilities, or certain sets of possibilities, beside and outside of which there exist others, and this obviously implies their limitation. In this instance still more can be said: to conceive of the Infinite quantitatively is not only to limit it, but in addition it is to conceive of it as subject to increase and decrease, which is no less absurd; with similar considerations one quickly finds oneself

---

5. In response to any astonishment that might arise on account of our use of the expression 'semi-profane', we will say that it is justified, in a very precise manner, by the distinction between effective initiation and merely virtual initiation, which we shall have to explain on another occasion. [See *Perspectives on Initiation*, chap. 30. ED.]

envisaging not only several infinites that coexist without confounding or excluding one another, but also infinites that are larger or smaller than others; and finally, the infinite having become so relative under these conditions that it no longer suffices, the 'transfinite' is invented, that is, the domain of quantities greater than the infinite. Here, indeed, it is properly a matter of 'invention', for such conceptions correspond to no reality. So many words, so many absurdities, even regarding simple, elementary logic, yet this does not prevent one from finding among those responsible some who even claim to be 'specialists' in logic, so great is the intellectual confusion of our times!

We should point out that just now we did not merely say 'to conceive of a quantitative infinite', but 'to conceive of the Infinite quantitatively', and this calls for a few words of explanation. By this expression we wanted to allude more particularly to those who are called 'infinitists' in contemporary philosophical jargon; indeed, all the discussions between 'finitists' and 'infinitists' clearly show that at least both have in common this completely false idea that the metaphysical Infinite is akin to the mathematical infinite, if they do not purely and simply identify the two.[6] Thus they all equally ignore the most elementary principles of metaphysics, since it is on the contrary precisely the conception of the true, metaphysical Infinite that alone allows us to reject absolutely every 'particular infinite', if one may so express it, such as the so-called quantitative infinite, and to be assured in advance that, wherever it is encountered, it can only be an illusion; we shall then only need to ask what could have brought about this illusion in order to be able to replace it with a notion closer to the truth. In short, every time it is a question of a particular thing, of a determined possibility, we can be certain *a priori* that it is limited by that very fact, and, we can say, limited by its very nature, and this holds equally true in the case where, for whatever

---

6. As a characteristic example, let us here cite the conclusion of L. Couturat's thesis *De l'infini mathématique*, in which he tried to prove the existence of an infinity of number and of magnitude by stating that his intention had been to show thereby that, 'in spite of neo-criticism [that is, the theories of Renouvier and his school], an infinitist metaphysics is plausible'!

reason, we cannot actually reach its limits; but it is precisely this impossibility of reaching the limits of certain things, and sometimes even of conceiving of them clearly, that causes the illusion that these things have no limits, at least among those for whom the metaphysical principle is lacking; and, let us say it again, it is this illusion and nothing more that is expressed in the contradictory assertion of a 'determined infinite'.

In order to rectify this false notion, or rather to replace it with a true conception of things,[7] we must here introduce the idea of the indefinite, which is precisely the idea of a development of possibilities the limits of which we cannot actually reach; and this is why we regard the distinction between the Infinite and the indefinite as fundamental to all questions in which the so-called mathematical infinite appears. Without doubt this is what corresponds in the intention of its authors to the Scholastic distinction between the *infinitum absolutum* and the *infinitum secundum quid*. It is certainly unfortunate that Leibniz, who had borrowed so much from Scholasticism, had neglected or not been aware of this, for however imperfect the form in which it was expressed, it would have allowed him to respond quite easily to certain objections raised against his method. In contrast to this, it seems that Descartes had indeed tried to establish the distinction in question, but he was very far from having expressed or even conceived of it with sufficient precision, since according to him the indefinite is that of which we do not perceive the limits, and which in reality could be infinite, although we could not affirm it to be so, whereas the truth is that we can on the contrary affirm that it is not so and that it is by no means necessary to perceive its limits in order to be certain that they exist. One can

---

7. One should, in all logical rigor, distinguish between a 'false notion' (or, if one prefer, 'pseudo-notion') and an 'incorrect notion'; an 'incorrect notion' is one that does not correspond adequately to reality, though it does, however, correspond in a certain measure; on the contrary, a 'false notion' is one that implies contradiction—as is the case here—and is therefore not really a notion, not even an incorrect one, though it appears as such to those who do not perceive the contradiction; for, expressing only the impossible, which is the same as nothingness, it corresponds to absolutely nothing; an 'incorrect notion' can be rectified, but a 'false notion' can only be rejected altogether.

11

thus see how vague and confused are all such explanations, and always as a result of the same lack of principle. Descartes indeed said: 'And for us, seeing things in which, in a certain sense,[8] we note no limits, we cannot ascertain thereby that they are infinite, but we shall only consider them to be indefinite.'[9] And he gives as examples the extension and divisibility of bodies; he does not contend that these things are infinite, but he does not seem to want to deny it formally either, and all the more so since he had just declared that he did not wish to 'entangle himself in disputes over the infinite,' which is rather too easy a way to brush aside the difficulties, even if he does say a little later that 'although we shall observe properties that seem to us not to have any limits, we do not fail to recognize that this proceeds from our lack of understanding and not from their nature.'[10] In short, he wishes with good reason to reserve the name infinite for what has no limits; but on the one hand he appears not to know with the absolute certitude that is implied in all metaphysical knowledge, that what has no limits cannot be anything but the universal All, and on the other hand the very notion of the indefinite needs to be much more precise; had it been so, a great number of subsequent confusions would doubtless not have been as readily produced.[11]

We say that the indefinite cannot be infinite because it always implies a certain determination, whether it is a question of extension, duration, divisibility, or some other possibility; in a word, whatever the indefinite may be, and according to whatever aspect it is considered, it is still of the finite and can only be of the finite. No

8. These words seem to refer to the Scholastic *secundum quid,* and thus it could be that the primary intention of the sentence cited had been to criticize indirectly the expression *infinitum secundum quid.*

9. *Principes de la Philosophie,* I, 26.

10. Ibid., I, 27.

11. Thus in his correspondence with Leibniz on the subject of the infinitesimal calculus, Varignon uses the terms 'infinite' and 'indefinite' indifferently, as if they were virtually synonymous, or at the very least as if it were unimportant, so to speak, that the one be taken for the other, even though it is on the contrary the difference in their meanings that should have been regarded as the essential point in all these discussions.

doubt, its limits may be extended until they are found to be out of our reach, at least insofar as we seek to reach them in a certain manner that we can call 'analytical', as we shall explain more thoroughly in what follows; but they are by no means abolished thereby, and in any case, if limitations of a certain order can be abolished, others possessing the same nature as the first will still remain, for it is by virtue of its nature, and not simply by some more or less exterior or accidental circumstances, that every particular thing is finite, whatever the degree to which certain limits can be extended. In this regard one might point out that the sign ∞, by which mathematicians represent their so-called infinite, is itself a closed figure, therefore visibly finite, just like the circle, which some people have wished to make a symbol of eternity, while it can in fact only be a figure of a temporal cycle, indefinite merely in its order, that is to say, of what is properly called perpetuity;[12] and it is easy to see that this confusion of eternity with perpetuity, so common among modern Westerners, is closely related to that of the Infinite and the indefinite.

In order to better understand the idea of the indefinite and the manner in which it is formed from the finite taken in its ordinary sense, one can consider an example such as that of the sequence of numbers: here, it is obviously never possible to stop at a determined point, since after every number there is always another that can be obtained by adding a unit; consequently, the limitation of this indefinite sequence must be of an order other than that which applies to a definite set of numbers taken between any two determined numbers; it must derive not from particular properties of certain numbers, but rather from the very nature of number in all its generality, that is to say from the determination that, essentially constituting this nature, makes number at once what it is and not anything else. One could make exactly the same observation if it were no longer a question of number but of space or time likewise

12. Again, we should note that, as we have explained elsewhere, such a cycle is never truly closed, and it seems so only so long as one places oneself in a perspective that does not allow one to perceive the distance really existing between its extremities, just as a helix situated along a vertical axis appears as a circle when it is projected on a horizontal plane.

considered in every possible extension to which they are subject.[13] Any such extension, as indefinite as one conceives it to be and as it in fact is, will never in any way take us out of the finite. Indeed, whereas the finite necessarily presupposes the Infinite—since the latter is that which comprehends and envelops all possibilities—the indefinite on the contrary proceeds from the finite, of which it is in reality only a development and to which it is consequently always reducible, for it is obvious that whatever process one might apply, one cannot derive from the finite either anything more or anything other than that which was already potentially contained therein. To take again the example of the sequence of numbers, we can say that this sequence, with all the indefinitude it implies, is given to us by its law of formation, since it is from this very law that its indefinitude immediately results; now this law consists in the following, that given any number, one can form the next by adding a unit. The sequence of numbers is therefore formed by successive additions of the unit to itself, indefinitely repeated, which is basically only the indefinite extension of the process of formation for any arithmetical sum; and here one can see quite clearly how the indefinite is formed starting from the finite. This example, moreover, owes its particular clarity to the discontinuous character of numerical quantity; but, to take things in a more general fashion applicable to all cases, it would suffice to insist on the idea of 'becoming' that is implied by the term 'indefinite', and this we expressed above in speaking of the development of possibilities, a development that in itself and in its whole course always consists of something unfinished;[14] the importance of the consideration of 'variables' as they concern the infinitesimal calculus will give to this last point its full significance.

13. It is thus of no use to say that space, for example, could be limited only by something still spatial, such that space in general could no longer be limited by anything; it is on the contrary limited by the very determination that constitutes its own nature as space and that leaves room, outside of it, to all the non-spatial possibilities.

14. Cf. the remark of A.K. Coomaraswamy on the Platonic concept of 'measure', which we have cited elsewhere (*The Reign of Quantity and the Signs of the Times*, chap. 3): the 'non-measured' is that which has not yet been defined, which is to say, in short, the indefinite, and it is at the same time and by the same token that which is only incompletely realized within manifestation.

2

The Contradiction
of Infinite Number

As we will see yet more clearly in the following, there are some cases
in which it suffices to replace the idea of the so-called infinite with
that of the indefinite in order to dispel all difficulties immediately;
but there are others in which even this is not possible, because it is a
question of something clearly determined—'fixed', so to speak, by
hypothesis—which, as such, cannot be called indefinite, according
to our last remarks above. Thus, for example, one can say that the
sequence of numbers is indefinite, but not that a certain number,
however great one supposes it to be and whatever position it occupies in the sequence, is indefinite. The idea of an 'infinite number',
understood as 'the greatest of all numbers', or 'the number of all
numbers', or, again, 'the number of all units', is in itself a truly contradictory idea, the impossibility of which would remain even were
one to renounce the unjustifiable use of the word 'infinite'. There
cannot be a number greater than all others, for however great a
number might be, one can always form a greater one from it by
adding a unit, in accordance with the law of formation which we set
forth above. This amounts to saying that the sequence of numbers
cannot have a final term, and it is precisely because it does not 'terminate' that it is truly indefinite; as the number of all the terms of
the sequence could itself only be the last of them, it can be said that
the sequence is not 'numerable', and this is an idea we shall have to
return to more fully in what follows.

The impossibility of an 'infinite number' can be established further by various arguments. Leibniz, who at least recognized this

quite clearly,[1] used one that consisted in comparing the sequence of even numbers to that of whole numbers: to every number there corresponds another number equal to its double, such that one can make the two sequences correspond term by term, with the result that the number of terms must be the same in both; but there are obviously twice as many whole numbers as there are even, since the even numbers alternate by twos in the sequence of whole numbers; one thus ends up with a manifest contradiction. One can generalize this argument by taking, instead of the sequence of even numbers, that is, multiples of two, that of multiples of any number whatsoever, and the reasoning will be identical; or again, in the same way one could take the sequence of the squares of whole numbers,[2] or more generally that of their powers of any exponent. Whatever the case, the conclusion will always be the same: a sequence containing only a part of the whole numbers will have the same number of terms as another containing all of them, which would amount to saying that the whole is not greater than its part, and, as soon as one allows that there is a number of all numbers, this contradiction will be inescapable. Nevertheless, some have thought to avoid it by supposing at the same time that there are numbers for which multiplication by a certain number or elevation to a certain power is not possible, precisely because such operations would yield a result exceeding the so-called 'infinite number'; there are even those who have indeed been led to envisage numbers said to be 'greater than infinite', whence such theories as that of Cantor's 'transfinite', which may be quite ingenious, but are no longer logically valid:[3] is it even

1. 'In spite of my infinitesimal calculus,' he wrote, 'I do not admit a true infinite number, though I do confess that the multitude of things surpasses all finite numbers, or rather all number.'

2. This was done by Cauchy, who attributed the argument, moreover, to Galileo (*Sept leçons de Physique générale*, third lesson).

3. Already at the time of Leibniz, Wallis was envisaging *spatia plus quam infinita* [more than infinite space]; this opinion, denounced by Varignon as implying contradiction, was equally held by Guido Grandi in his book *De Infinitis infinitorum* [Concerning the Infinite of infinites]. On the other hand, Jean Bernoulli, in the course of his discussions with Leibniz, wrote, *Si dantur termini infiniti, dabitur etiam terminus infinitesimus (non dico ultimus) et qui eum equuntur* [If the limits of

16

conceivable that one could dream of calling a number 'infinite' when it is on the contrary so 'finite' that it is not even the greatest of all numbers? Moreover, with such theories there would be numbers to which none of the rules of ordinary calculation would apply any longer, or, in short, numbers that would no longer truly be numbers but merely called such by convention.[4] This inevitably occurs when, seeking to conceive of an 'infinite number' otherwise than as the greatest of all numbers, one envisages various 'infinite numbers', supposedly unequal to each other, to which we attribute properties that no longer have anything in common with those of ordinary numbers; thus one escapes one contradiction only to fall into others, and all this is at bottom only the product of the most meaningless 'conventionalism' imaginable.

Thus, the idea of a so-called 'infinite number', whatever manner in which it is presented and whatever name by which one wishes to designate it, always comprises contradictory elements; moreover, one has no need of such an absurd supposition from the moment one forms a proper conception of what the indefinitude of number really is, and when one further recognizes that number, despite its indefinitude, is by no means applicable to all that exists. We need not dwell upon this last point here, as we have already sufficiently explained it elsewhere. Number is only a mode of quantity, and quantity itself only a category or special mode of being, not coextensive with it, or, more precisely still, quantity is only a condition proper to one certain state of existence in the totality of universal existence; but this is precisely the point that most moderns have difficulty understanding, habituated as they are to wanting to reduce everything to quantity and even to evaluating everything

the infinite are given, the infinitesimal limits will also be given (I do not say the ultimate limits) which follow upon them], which, though he never explained himself more clearly, seems to indicate that he supposed that in a numerical sequence there could be terms 'beyond the infinite'.

4. One can by no means say that here it is a question of an analogical use of the idea of number, for this would imply transposition to a domain other than that of quantity; on the contrary, considerations of this sort always refer exclusively to quantity understood in its most literal sense.

numerically.[5] However, in the very domain of quantity there are things that escape number, as we shall see when we come to the subject of continuity; and even without departing from the sole consideration of discontinuous quantity, one is already forced to admit, at least implicitly, that number is not applicable to everything, when one recognizes that the multitude of all numbers cannot constitute a number, which, moreover, is finally only an application of the incontestable truth that what limits a certain order of possibilities must necessarily be beyond and outside that which it limits.[6] Only it must be understood that such a multitude, be it discontinuous, as in the case of the sequence of numbers, or continuous—a subject we shall have to return to shortly—can in no wise be called infinite, and in such cases there can never be anything but the indefinite; and it is this notion of multitude that we are now going to examine more closely.

5. Thus Renouvier thought that number is applicable to everything, at least ideally, that is, that everything is 'numerable' in itself, even if we are in fact incapable of 'numbering' it; he therefore completely misunderstood the meaning Leibniz gives to the notion of 'multitude', and he was never able to understand how the distinction between the latter and number allows one to escape the contradiction of an 'infinite number'.

6. We have said, however, that every particular or determined thing, whatever it might be, is limited by its very nature, but there is absolutely no contradiction here: indeed, it is limited by the negative side of this nature (for, as Spinoza has said, *omnis determinatio negatio est* [all determination is a negation]), that is, its nature considered insofar as it excludes other things and leaves them outside of itself, so that finally it is really the coexistence of these other things that limits the thing in consideration; this is moreover why the universal All, and it alone, cannot be limited by anything.

3

The Innumerable Multitude

As we have seen, Leibniz by no means admits 'infinite number', since on the contrary he expressly declares that this would imply contradiction in whatever sense one took it; on the other hand, he does admit what he calls an 'infinite multitude', though without making it clear—as the Scholastics would at least have done—that in any case it can only be an *infinitum secundum quid*, the sequence of numbers being, for him, an example of such a multitude. From another point of view, however, in the quantitative domain, and even in that of continuous magnitude, the idea of the infinite always appears to him as suspect of at least possible contradiction, for, far from being an adequate idea, it inevitably entails a certain amount of confusion, and we cannot be certain that an idea implies no contradiction unless we distinctly conceive all of its elements;[1] this hardly allows according this idea a 'symbolic'—we would rather say 'representative'—character, and as we shall see later, this is why he

---

1. Descartes spoke solely of 'clear and distinct' ideas; Leibniz specified that an idea can be clear without being distinct, in that it only allows one to recognize it and to distinguish it from all other things, whereas a distinct idea is that which is not only 'distinguishing' in this sense, but 'distinguished' in its elements; moreover, an idea can be more or less distinct, and the adequate idea is that which is so completely and in all its elements; but, while Descartes was of the opinion that one could have 'clear and distinct' ideas of all things, Leibniz on the contrary believed that mathematical ideas alone can be adequate, their elements being as it were of a definite number, whereas all other ideas enclose a multitude of elements, of which the analysis can never be completed, so that they will always remain partially confused.

never dared to give a clear verdict on the reality of the 'infinitely small'; but this very perplexity, this doubtful attitude, brings out even better the lack of principle that led him to admit that one could speak of an 'infinite multitude'. From this one might also wonder if he did not think that in order to be 'infinite', as he calls it, such a multitude must not only be 'numerable', which is obvious, but that it must not even be quantitative at all, taking quantity in all its extension and in all its modes; this would be true in certain cases, but not in all; however it may be, it remains a point on which he never clearly explained himself.

The idea of a multitude that surpasses all number, and that consequently is not a number, seems to have astonished most of those who have discussed the conceptions of Leibniz, be they 'finitists' or 'infinitists'; it is nevertheless far from unique to Leibniz, as they have generally seemed to believe, and, on the contrary, was quite common among the Scholastics.[2] This idea was applied specifically to everything that is neither a number nor 'numerable', that is, all that does not relate to the domain of discontinuous quantity, whether it be a question of things belonging to other modes of quantity, or of what is entirely outside of the quantitative domain, for it concerned an idea belonging to the order of 'transcendentals', or general modes of being, which, contrary to its special modes like quantity, are coextensive with it.[3] This also allows one to speak of the multitude of divine attributes for example, or again of the multitude of angels, that is, of beings belonging to states that are not

2. We will cite only one text among others, which is particularly clear in this regard: *Qui diceret aliquam multitudinem esse infinitam, non diceret eam esse numerum, vel numerum habere; addit etiam numerus super multitudinem rationem mensurationis. Est enim numerus multitudo mensurata per unum ... et propter hoc numerus ponitur species quantitatis discretae, non autem multitudo, sed est de transcendentibus* [If one were to say that some multitude is infinite one would not be saying that it is a number or has a number, for number adds to multitude the idea of measure. For a number is multitude measured by one ... and for this reason number is categorized as a species of discrete quantity but multitude is not, but rather is one of the transcendentals (Saint Thomas Aquinas, in *Physics*, III, 1.8).

3. We know that the Scholastics, even in the properly metaphysical part of their doctrines, never went beyond the consideration of Being, so that for them metaphysics is in fact reduced solely to ontology.

subject to quantity, where, consequently, there can be no question of number; it is also this that allows one to speak of the states of being or degrees of existence as multiple or as constituting an indefinite multitude, even though quantity is only one special condition of a single one of them. On the other hand, since the idea of multitude, contrary to that of number, is applicable to all that exists, there must necessarily be multitudes of a quantitative order, notably in the domain of continuous quantity, and this is why we said just now that it would not be correct to consider every case of the so-called 'infinite multitude', that is, that which surpasses all number, as entirely escaping the domain of quantity. Furthermore, number itself can also be regarded as a species of multitude, but on the added condition that it be a 'multitude measured by the unit', according to the expression of Saint Thomas Aquinas; all other sorts of multitude, being 'innumerable', are 'non-measured', which is not to say they are infinite, but merely that they are indefinite.

While on the subject, it is appropriate to note a rather singular fact: for Leibniz, this multitude, which does not constitute a number, is nonetheless a 'result of units'.[4] How should we understand this, and indeed, what are the units in question? The word unit can be taken in two completely different senses:[5] on the one hand, there is the arithmetical or quantitative unit, which is the first element of number, its point of departure, and, on the other hand, there is what is analogously designated as metaphysical Unity, which is identified with pure Being itself; we see no other possible meaning outside of these; but furthermore, whenever one speaks of 'units' in the plural, this can obviously only be understood in the quantitative sense. If this is so, however, then the sum of these units cannot be anything other than a number, and can in no way transcend number; it is true that Leibniz said 'result' and not 'sum', but this distinction, even if it is intentional, nonetheless remains an unfortunate obscurity. Besides, he declares elsewhere that multitude, without being a number, is nevertheless conceived by analogy with

4. *Système nouveau de la nature et de la communication des substances.*
5. The French word *unité* means both 'unit' and 'unity', as Guénon himself explains. ED.

21

number: 'When there are more things,' he says, 'than can be comprehended by any number, we yet attribute to them analogically a number that we call infinite,' although this would only be a 'manner of speaking', a *modus loquendi*,[6] and even, in this form, a most incorrect manner of speaking, since in reality the thing in question is not a number at all; but whatever the imperfections of expression and the confusions to which they might give rise, we must in any case admit that an identification of multitude with number was assuredly not at the root of his thought.

Another point to which Leibniz seems to attach great importance is that the 'infinite', such as he conceives of it, does not constitute a whole;[7] this is a condition he regards as necessary if the idea is to escape contradiction, but here we have another rather obscure point. One might well wonder what sort of 'whole' is in question here, and it is first of all necessary to put aside entirely the idea of the universal All, which is on the contrary, as we have said from the beginning, the metaphysical Infinite itself, the only true Infinite, which could by no means be in question here; indeed, whether it is a question of continuous or discontinuous, the 'indefinite multitude' that Leibniz envisages in any case only makes sense in a restricted and contingent domain of a cosmological and not metaphysical order. It is obviously a question, moreover, of a whole conceived as composed of parts, whereas, as we have explained elsewhere,[8] the universal All is properly 'without parts', by very reason of its infinity, since these parts are necessarily relative and finite and thus could

6. *Observatio quod rationes sive proportiones non habeant locum circa quantitates nihilo minores, et de vero sensu Methodi infinitesimalis* [An Observation that Calculations and Proportions Do Not Apply to Diminishing Quantities, and About the True Understanding of the Infinitesimal Method], in the *Acta Eruditorum* of Leipzig, 1712.

7. Cf. ibid., *Infinitum continuum vel discretum proprie nec unum, nec totum, nec quantum est* [The continuous or discrete infinite is properly speaking neither one nor a whole nor a quantity], where the expression *nec quantum* seems to imply that for him, as we indicated above, the 'indefinite multitude' must not be conceived of quantitatively, unless by *quantum* he had meant solely a definite quantity, as the so-called 'infinite number' would have been, the contradiction of which he had already demonstrated.

8. On this point, see further *The Multiple States of the Being*, chap. 1.

not have any real connection with it, which amounts to saying that for it they do not exist. So, as regards the question posed, we must confine ourselves to the consideration of a particular whole; but here again, and precisely in what concerns the mode of composition of such a whole and its relation with its parts, there are two cases to consider, corresponding to two very different senses of the same word 'whole'. First, there is the whole that is nothing more or other than the simple sum of its parts, of which it is composed in the manner of an arithmetical sum, which Leibniz says is obviously fundamental, for this mode of formation is precisely that which is proper to number, and he does not allow us to go beyond number; but in fact this notion, far from representing the only way in which a whole can be conceived, is not even that of a true whole in the most rigorous sense of the term. Indeed, a whole that is thus only the sum or result of its parts and which consequently is logically posterior to them, is, as such, nothing other than an *ens rationis* [a being of reason or of the mind], for it is 'one' and 'whole' only in the measure that we conceive it as such; in itself it is strictly speaking only a 'collection', and it is we who, by the manner in which we envisage it, confer upon it in a certain relative sense the character of unity and totality. On the contrary, a true whole possessing this character by its very nature, must be logically anterior to its parts and independent of them: such is the case with a continuous set, which we can divide into parts arbitrarily, that is, into parts of any size, without in the least presupposing the actual existence of these parts; here, it is we who give a reality to the parts as such, by an ideal or effective division, and this case is thus the exact inverse of the preceding.

Now, the whole question comes back in short to knowing whether, when Leibniz says that 'the infinite is not a whole,' he excludes this second sense as well as the first; it seems that he does, and this is probable since it is the only case in which a whole would truly be 'one', and since the infinite, according to him, is *nec unum, nec totum* [neither one nor a whole]. What further confirms this is that this latter, and not the former, is what applies to a living being or an organism when it is considered from the point of view of totality; now Leibniz says: 'Even the Universe is not a whole, and it must not be conceived of as an animal with God for its soul, as the

23

ancients thought.'[9] However, if this is so, one does not really see how the ideas of the infinite and the continuous can be connected, as he most often takes them to be, since the idea of the continuous is, at least in a certain sense, linked precisely to this second conception of totality; but this is a point that will be better understood in the light of what is to follow. In any case, what is certain is that if Leibniz had conceived of a third sense of the word 'whole', a purely metaphysical sense superior to the other two, namely the idea of the universal All as we set it forth at the very beginning, he would not have been able to say that the idea of the infinite excludes totality, for he declares moreover: 'The real infinite is perhaps the absolute itself, which is not composed of parts, but having parts, comprehends them by eminent reason, as to the degree of its perfection.'[10] Here, one could say, there is at the very least a 'glimmer', for this time, almost by exception, he takes the word 'infinite' in its true sense, although it would be erroneous to say that this infinite 'has parts', however one wishes to understand this; but it is then strange that he again expresses his thought only in a doubtful and perplexing form, as if he were not exactly settled as to the significance of the idea; and indeed perhaps he never was, for otherwise one could not explain why he so often turned away from its proper meaning, and why, when he speaks of the infinite, it is sometimes so difficult to know whether his intention was to take this term rigorously, albeit wrongly, or whether he had in view only a simple 'manner of speaking'.

9. Letter to Jean Bernoulli.—Leibniz here rather gratuitously attributes to the ancients in general an opinion that in reality was held by only some of them; he obviously had in mind the theory of the Stoics, who conceived of God as uniquely immanent, identifying him with the *Anima Mundi*. It goes without saying, moreover, that it is here a question only of the manifested Universe, that is, the cosmos, and not of the universal All, which comprehends all possibilities, the non-manifested as well as the manifested.

10. Letter to Jean Bernoulli, June 7, 1698.

4

The Measurement
of the Continuous

Until now, when speaking of number we have had in view whole number exclusively,[1] and logically this was so of necessity, since we were regarding numerical quantity strictly as discontinuous quantity: between two consecutive terms in the sequence of whole numbers there is always a perfectly definite interval, marked by the difference of a unit existing between these two numbers, which, when one keeps to the consideration of whole number, is in no way reducible. In reality, moreover, it is whole number alone that is true number, what one might call pure number; and the sequence of whole numbers, starting from the unit, continues increasing indefinitely without ever arriving at a final term, the supposition of which, as we have seen, would be contradictory; but it goes without saying that the sequence develops entirely in a single direction, and so the other, opposite direction—that of indefinite decrease— cannot be represented by it, although from another point of view there is a certain correlation and a sort of symmetry between the considerations of indefinitely increasing and indefinitely decreasing quantities, as we shall demonstrate further on. However, people have not stopped at whole number, but have been led to consider various kinds of number; it is usually said that these are extensions

---

1. 'Whole numbers' (*nombres entiers*) is simply what is nowadays termed 'integers', which is to say that the term 'whole number' (even though everyone will immediately understand what is meant) is not currently idiomatic. It appears, moreover, that when Guénon speaks of *nombres entiers*, he means the positive integers, or so-called natural numbers. ED.

or generalizations of the idea of number, and this is true after a certain fashion; but at the same time these extensions are also distortions, and this modern mathematicians seem too easily to forget, since their 'conventionalism' leads them to misunderstand the origin and raison d'être of these numbers. In fact, numbers other than whole numbers always appear above all as the representation of the results of operations that would be impossible were one to keep to the point of view of pure arithmetic, which in all rigor is the arithmetic of whole numbers alone: thus a fractional number, for example, is no more than the representation of the result of a division that cannot in fact be made, that is, one that must be declared arithmetically impossible, and this, moreover, is implicitly recognized when it is said, according to ordinary mathematical terminology, that one of the two numbers in question is not divisible by the other. Here we should point out that the definition commonly given to fractional numbers is absurd; fractions can in no way be 'parts of a unit', as is said, for the true arithmetical unit is necessarily indivisible and without parts; and from this results the essential discontinuity of number, which is formed from the unit; but let us see whence this absurdity arises.

Indeed, one does not arbitrarily consider the results of the aforementioned operations thus, instead of regarding them purely and simply as impossible; generally speaking, it is in consequence of the application made of number—discontinuous quantity—to the measurement of magnitudes belonging to the order of continuous quantity, as, for example, spatial magnitudes. Between these two modes of quantity is a difference of nature such that a correspondence between the two cannot be perfectly established; to remedy this to a certain point, at least insofar as it is possible, one seeks to reduce, as it were, the intervals of this discontinuity constituted by the sequence of whole numbers, by introducing other numbers between its terms, and fractional numbers first of all, which would be meaningless apart from this consideration. It is then easy to understand that the absurdity we just pointed out concerning the definition of fractions arises quite simply from a confusion of the arithmetical unit with what are called 'units of measurement', units that are such only by convention, and that in reality are magnitudes of another sort than number, notably geometric magnitudes. The unit of length, for example, is only a certain length chosen for reasons foreign to arithmetic, and the number 1 is made to correspond to it in order to be able to measure all other lengths by reference thereto; but all length, even when so represented by the unit, is by its very nature as continuous magnitude no less always and indefinitely divisible. Comparing it to other lengths that are not exact multiples of it, one might thus have to consider parts of this unit of measurement, which would in no way be parts of the arithmetical unit on that account; and it is only thus that the consideration of fractional numbers is really introduced, as a representation of the ratios of magnitudes that are not exactly divisible by one another. The measurement of a magnitude is indeed no more than the numerical expression of its ratio to another magnitude of the same kind taken as the unit of measurement, or, basically, as the term of comparison; and this is why the ordinary method of measuring geometric magnitudes is essentially founded on division.

It must be said, moreover, that in spite of this method something of the discontinuous nature of number is always bound to remain, preventing one from thus obtaining a perfect equivalent to the continuous; reduce the intervals as much as one likes—which finally is to say, reduce them indefinitely, rendering them smaller than any quantity that can be given in advance—but they will never be done away with entirely. To make this clearer, let us take the simplest example of a geometric continuum, a straight line: we shall consider half a straight line, extending indefinitely in a certain direction,[2] and let us agree to make each of its points correspond to a number expressing the distance of the point from the origin, represented by zero, as its distance from itself is obviously nothing; starting from this origin, the whole numbers will then correspond to the successive extremities of all segments equal to each other and to the unit of length; the points contained between these will be representable

---

2. It will be seen in what follows, concerning the geometric representation of negative numbers, why we must take into consideration here only half a straight line; besides, the fact that the series of numbers develops only in a single direction, as we said earlier, should already suffice to indicate the reason.

only by fractional numbers, since their distances from the origin are not exact multiples of the unit of length. It goes without saying that, taking fractional numbers with greater and greater denominators, hence smaller and smaller differences, the intervals between the points to which these numbers correspond will be reduced in the same proportion; in this way the intervals can be decreased indefinitely, theoretically at any rate, since the possible denominators of the fractional numbers are themselves whole numbers, the sequence of which increases indefinitely.[3] We say theoretically because in fact the multitude of fractional numbers is indefinite, and one could never use them all, but let us suppose that ideally all the possible fractional numbers could be made to correspond to the points on the half of the line in consideration. Despite the indefinite decrease of the intervals, a multitude of points to which no number will correspond will still remain on this line. At first this might seem strange and even paradoxical, but it is nevertheless easily demonstrated, for such a point can be obtained by means of a very simple geometric construction. Let us construct a square having for its side the line segment with extremities at the points 0 and 1, and let us draw the diagonal of the square starting from the origin, then a circle having for its center the origin and for its radius this diagonal; the point at which this circle cuts the straight line cannot be represented by any whole or fractional number, since its distance from the origin is equal to the diagonal of the square, which is incommensurable with its side, that is, with the unit of length. Thus, the multitude of fractional numbers, despite an indefinite decrease of their differences, still does not suffice to fill, so to speak, the intervals between the points contained in the line,[4] which amounts to saying that this multitude is not a real and adequate equivalent to linear continuity; in order to express the measurement of certain lengths, one is thus forced to introduce still other kinds of numbers, what are called incommensurable numbers, that is, those having no

---

3. This will be made still clearer when we come to speak of negative numbers.

4. Note that we did not say the points composing or constituting the line, which would betray a false understanding of continuity, as considerations we shall later explain will show.

common measure with the unit. Such are the irrational numbers, which represent the results of arithmetically impossible extractions of roots, as, for example, the square root of a number that is not a perfect square; thus in the preceding example, the ratio of the diagonal of the square to its side, and consequently the point having a distance from the origin equal to this diagonal, can be represented only by the irrational number $\sqrt{2}$, which is indeed incommensurable, for there exists no whole or fractional number the square of which is equal to 2; and besides these irrational numbers there are still other incommensurable numbers, the geometrical origin of which is obvious, as, for example, the number π, which represents the ratio of the circumference of a circle to its diameter.

Without entering further into the question of the 'composition of the continuous', it will thus be seen that number, however far the notion is extended, is never perfectly applicable to it; finally this application always amounts to replacing the continuous with a discontinuity, the intervals of which can be very small, and can even become smaller and smaller still by an indefinite series of successive divisions, but without ever being done away with, for in reality there is no 'final term' to which the divisions might lead, since a continuous quantity, however small it might be, will always remain indefinitely divisible. It is to these divisions of the continuous that the consideration of fractional numbers properly corresponds; but, and this is particularly important to note, a fraction, however minute it might be, is always a determined quantity, and however small one supposes the difference between two fractions there is always an equally determined interval. Now the property of indefinite divisibility that characterizes continuous magnitudes obviously demands that one always be able to take elements as small as one wishes, and that the intervals existing between these elements can likewise be rendered less than any given quantity; but—and it is here that we see the insufficiency of fractional numbers, and even, we can say, of number altogether—in order that there really be continuity, these elements and these intervals must not be conceived of as something determined. Consequently, the most perfect representation of continuous quantity will be obtained by the consideration not of fixed and determined magnitudes such as those just discussed, but on the contrary of variables, for then their variability can itself be regarded as accomplished in a continuous fashion; and these quantities must be capable of indefinite decrease by virtue of their variability, without ever canceling themselves out or reaching a 'minimum', which would be no less contradictory than 'final terms' of the continuous: here, precisely, as we shall see, is the true notion of infinitesimal quantities.

5

Questions
Raised by the
Infinitesimal Method

When Leibniz first presented the infinitesimal method,[1] and even
again in several other works that followed,[2] he particularly emphasized the uses and applications of the new calculus, in keeping with
the modern tendency to attribute more importance to the practical
applications of science than to science itself, as such; it would be difficult to say whether this tendency truly existed in Leibniz, or
whether this manner of presenting his method was only a sort of
concession on his part. Be that as it may, in order to justify a
method it certainly does not suffice to show the advantages it might
have over other, previously accepted methods, or the conveniences
it might furnish practically for calculation, nor even the results it
might in fact have given; and the adversaries of the infinitesimal
method did not fail to make use of this, and it was only their objections that persuaded Leibniz to explain the principles, and even the
origins, of his method. It is very possible, moreover, that on this last
point he might never have spoken at all, but ultimately this is of little importance, for very often the occasional causes of a discovery

1. *Nova Methodus pro maximis et minimis, itemque tangentibus, quae nec fractas
nec irrationales quantitates moratur, et singulare pro illis calculi genus* [A New
Method for Greatest and Smallest Quantities as Well as Tangents, Which Does Not
Involve Either Fractional or Irrational Quantities, and a Unique Kind of Calculus
For Them], in the *Acta Eruditorum* of Leipzig, 1684.

2. *De Geometria recondita et Analysi indivisibilium atque infinitorum* [On the
Hidden Geometry and the Analysis of Indivisible and Infinite Quantities], 1686.
Subsequent works all relate to the solving of particular problems.

are in themselves only rather insignificant circumstances; at any rate, of what he wrote on the subject,[3] all that interests us is the fact that he passed from a consideration of the 'assignable' differences existing between numbers to a consideration of the 'unassignable' differences that can be conceived of between geometric magnitudes by reason of their continuity, and that he also attached great importance to this order, as being so to speak 'demanded by the nature of things'. From this it follows that for him infinitesimal quantities do not naturally appear directly to us, but only as a result of passing from a consideration of the variability of discontinuous quantity to that of continuous quantity, and from the application of the first to the measurement of the second.

What exactly is the meaning of these infinitesimal quantities Leibniz was reproached for using without having first defined what he meant by them, and did this meaning allow him to regard his calculus as absolutely rigorous, or on the contrary merely as a method of approximation? To respond to these two questions would, by that very fact, be to resolve the most important objections raised against him; but unfortunately he himself never responded very clearly, and even his various attempts to do so do not always seem in complete accord with one another. In this connection it is worth noting that generally speaking Leibniz was in the habit of explaining the same thing differently according to the audience he was addressing; we would certainly not hold this behavior against him, which is irritating only for systematic minds, for in principle he was only conforming to an initiatic and, more particularly, Rosicrucian precept according to which it is fitting to speak to each in his own language; only he sometimes happened to apply the precept rather poorly. Indeed, if it is obviously possible to clothe the same truth in different expressions, it is understood that this be done without ever distorting or diminishing it, being always careful to refrain from any manner of speaking that could give rise to false conceptions; in this regard Leibniz failed in a number of instances.[4] Thus, he pushed the

3. First in his correspondence, and then in *Historia et origo Calculi differntialis* [The History and Origin of Differential Calculus], 1714.

4. In Rosicrucian language one would say that this, as much as and even more than the failure of his projects of *characteristica universalis*, proves that even if he

idea of 'accommodation' to the point of sometimes seeming to justify those who wished to see in his calculus merely a method of
approximation, for at times he presented it as being no more than a
sort of abridged version of the ancients' 'method of exhaustion', useful for facilitating calculations but yielding results that have to be
verified by this other method if a rigorous demonstration is desired;
and it is nevertheless quite certain that this was not fundamentally
what he thought, but that, in reality, he saw in it much more than a
simple expedient intended to shorten calculations.

Leibniz frequently declared that infinitesimal quantities cannot
but be 'incomparable', but as to the precise meaning in which this
word is to be understood, he gave an explanation that is not only
rather unsatisfying, but even most regrettable, for it could not but
provide ammunition to his adversaries, who, moreover, did not fail
to avail themselves of it; here, again, he was certainly not expressing
what he truly thought, and we can see in this another example of an
excessive 'accommodation', yet more serious than the first, that
would substitute erroneous views for 'adapted' expressions of the
truth. Leibniz writes:

One need not take the infinite here rigorously, but only in the
manner in which one says in optics that the rays of the sun come
from an infinitely distant point, and may thus be treated as
parallel. And when there are several degrees of the infinite or of
the infinitely small, this is like the terrestrial globe being regarded
as a point with respect to the distance of the fixed stars, and a ball
we might take in hand being again a point in comparison with
the semi-diameter of the terrestrial globe, such that the distance
of the fixed stars is like an infinite infinitude with respect to the
diameter of the ball. For instead of the infinite or the infinitely
small, one takes quantities as great or as small as is necessary for
the error to be less than a given error, such that one differs from
the style of Archimedes only in expression, which in our method
is more direct, and more conformable with the art of invention.[5]

did have some theoretical idea of the nature of the 'gift of tongues', he was nevertheless far from having received it effectively.

5. 'Mémoire de M.G.G. Leibniz touchant son sentiment sur le Calcul differen-
tiel', in the *Journal de Trévoux*, 1701.

It was unfailingly pointed out to Leibniz that however small the terrestrial globe might be with respect to the heavens, or a grain of sand in relation to the terrestrial globe, they are nonetheless fixed and determined quantities, and if one of these quantities can be regarded as practically negligible in comparison with the other, this is nevertheless only a simple approximation; his reply was that he had only wished to 'avoid the subtleties' and to 'make the reasoning evident to all,'[6] which fully confirms our interpretation, and which, furthermore, is already a sort of manifestation of the 'popularizing' tendency of modern scholars. What is most extraordinary is that he was able to write afterwards: 'At any rate, there was not the slightest thing that should have caused anyone to imagine that I indeed meant a very small, but always fixed and determined, quantity,' to which he added: 'Besides, I had already written some years ago to Bernoulli of Groningen that the infinites and infinitely small might be taken for fictions, similar to imaginary roots,[7] without thereby harming our calculus, these fictions being useful and founded in reality.'[8] Moreover, it seems that he never did understand exactly in what respect his comparison was flawed, for he presents it again in the same terms about ten years later;[9] but, at any rate since he expressly declared that his intention had not been to present the infinitesimal quantities as determined, we must conclude from this that, for him, the meaning of the comparison amounts to the following: a grain of sand, though not infinitely small, can, however, without appreciable disadvantage, be considered as such in relation to the earth, and thus there is no need to envisage the infinitely small 'rigorously'—they may even be regarded as mere fictions if one so desires; but however one takes them, such a consideration is nonetheless manifestly unsuitable to give any other idea of the infinitesimal calculus than that of a simple calculus of approximation, which would assuredly have been insufficient in the eyes of Leibniz himself.

6. Letter to Varignon, February 2, 1702.

7. Imaginary roots are roots of negative numbers; later we shall speak more of the question of negative numbers and the logical difficulties to which they give rise.

8. Letter to Varignon, April 14, 1702.

9. Mémoire already cited above, in the *Acta Eruditorum* of Leipzig, 1712.

6

Well-Founded
Fictions

The thought most characteristic of Leibniz, although he does not always affirm it with the same force, and on which he sometimes even seems, albeit exceptionally, not to wish to deliver a categorical verdict, is that basically infinite and infinitely small quantities are only fictions; but, he adds, they are 'well-founded fictions', and by this he does not simply mean that they are useful for calculation,[1] or even for 'finding real truths', although sometimes he does also insist on this usefulness; but he constantly repeats that these fictions are 'founded in reality', that they are *fundamentum in re*, which obviously implies something of a more than purely utilitarian value; and for him this value itself must after all be explained by the basis these fictions have in reality. In any case, he believes that for the method to be reliable, it suffices to envisage, not infinite and infinitely small quantities in the rigorous sense of these expressions, since this would have no corresponding reality, but simply quantities as great or as small as one likes, or as is necessary in order for the error to be rendered less than any given quantity. It is still necessary to examine whether it is true that, as he declares, this error is thereby null, that is, whether this manner of envisaging the infinitesimal calculus gives him a perfectly rigorous foundation, but we shall have to return to this question later. However it might be with

---

1. It is in this consideration of practical utility that Carnot believed he had found a sufficient justification; it is obvious that from the time of Leibniz to him, the 'pragmatist' tendency of modern science had already become much more pronounced.

respect to this last point, for him statements concerned with the infinite and infinitely small quantities fall under the category of assertions that according to him are only *toleranter verae* [reasonably true], or 'tolerable', and must be 'redressed' by an explanation, as when one regards negative quantities as 'less than zero', or as in a number of other cases in which the language of geometry implies 'a certain figurative and cryptic manner of speaking';[2] the word 'cryptic' would seem to be an allusion to the symbolic and profound meaning of geometry, but this is not at all what Leibniz had in mind, and perhaps as is so often the case with him in so speaking he had only the memory of some esoteric notion, more or less poorly understood.

As for the sense in which one should understand the statement that infinitesimal quantities are 'well-founded fictions', Leibniz declared that 'the infinites and infinitely small are founded in such a way that within the realm of geometry, and even in nature, they may be treated as if they were perfectly real';[3] indeed, for him, everything that exists in nature in some way implies the consideration of the infinite, or at least of what he believed could be called such. As he said, 'the perfection of the analysis of transcendentals, or of geometry involving the consideration of some infinite would without doubt be all the more important on account of the applications one can make of it to the operations of nature, which introduces the infinite in all that it does';[4] but perhaps this is only because we cannot have adequate ideas of it, and because it always introduces elements we cannot perceive with complete distinctness. If this is so, then it is necessary not to take too literally such assertions as the following for example: 'Since our method is properly that part of general mathematics that treats of the infinite, one has great need of it in applying mathematics to physics, for as a rule the character of the infinite Author enters into the operations of nature.'[5] But if by

---

2. Previously cited Mémoire, in the *Acta Eruditorum* of Leipzig, 1712.
3. Previously cited letter to Varignon, February 2, 1702.
4. Letter to Marquis de l'Hospital, 1693.
5. 'Considérations sur la différence qu'il y a entre l'Analyse ordinaire et le nouveau Calcul des transcendantes', in the *Journal des Sçavans*, 1694.

this even Leibniz only means that the complexity of natural things goes incomparably beyond the limits of distinct perception, it nonetheless remains that the infinite and infinitely small quantities must have their *fundamentum in re*; and this foundation is found in the nature of things, at least as conceived by him, and is none other than what he calls the 'law of continuity', which we shall have to examine a little later, and which he regards, rightly or wrongly, as being in short only a particular case of a certain 'law of justice', which is itself ultimately connected to the idea of order and harmony, and which equally finds its application every time a certain symmetry must be observed, as, for example, in the case of combinations and permutations.

Now, if the infinite and infinitely small quantities are only fictions, and even admitting that they really are 'well-founded', one might ask oneself this: why use such expressions, which, even if they can be regarded as *toleranter verae*, are nonetheless incorrect? Here is something which presages, one might say, the 'conventionalism' of modern science, though with the notable difference that the latter is no longer in any way preoccupied with knowing whether the fictions to which it has recourse are 'well-founded' or not, or, according to another expression of Leibniz, whether they can be interpreted *sano sensu* [in a reasonable way], or even whether they have any meaning at all. Moreover, since one can do without these fictional quantities and be content with envisaging in their place quantities that can simply be rendered as great or as small as one likes, and which, for that reason, can be said to be indefinitely great or indefinitely small, it would no doubt have been better to do so from the start and thus avoid introducing fictions that, whatever might be their *fundamentum in re*, are, ultimately, of no practical use, not only with regard to calculation, but even regarding the infinitesimal method itself. The expressions 'indefinitely great' and 'indefinitely small', or what amounts to the same but is perhaps more precise, 'indefinitely increasing' and 'indefinitely decreasing', not only have the advantage of being the only ones that are rigorously exact; they also show clearly that the quantities to which they are applied can only be variable, and not determined, quantities. As a mathematician has rightly said, 'the infinitely small is not a very small quantity, having an actual value capable of being determined; its character is to be eminently variable, and to be able to take on a value less than that of any other one might wish to specify; it would be much better to call them indefinitely small.'[6]

The use of these terms would have prevented many difficulties and disputes, and there is nothing surprising about this, since it is not a simple question of words, but the replacement of a false idea with a true one, of a fiction with a reality; notably, it would have prevented anyone from taking the infinitesimal quantities to be fixed and determined quantities, for as we said above the word 'indefinite' always carries with it the idea of 'becoming', and consequently of change, or, when it is a question of quantities, of variability; and, had Leibniz made a habit of using these terms, he would doubtless not have allowed himself to be so easily drawn into the unfortunate comparison concerning the grain of sand. What is more, reducing *infinite parva ad indefinite parva* [the infinitely small to the indefinitely small] would at any rate have been clearer than reducing them *ad incomparabiliter parva* [to the incomparably small]; precision would thereby have been gained without any loss of exactitude—quite the contrary. Infinitesimal quantities assuredly are 'not comparable' to ordinary quantities, but this can be understood in more than one way, and indeed it has often enough been taken in other senses than were intended. It is better to say that they are 'unassignable', to use another expression of Leibniz, for it seems that this term can be rigorously understood only of quantities that are capable of becoming as small as one likes, that is, smaller than any given quantity, and consequently to which one can by no means 'assign' a determined value, however small it might be, and this is indeed the sense of *indefinite parva*. Unfortunately, it is next to impossible to know whether, in Leibniz's thought, 'incomparable'

6. Ch. de Freycinet, *De l'Analyse infinitésimale*, pp 21–22. The author adds: 'But the first expression [that of infinitely small] having prevailed in the language, we believe it should be retained.' This is assuredly quite an excessive scruple, for usage does not suffice to justify the mistakes and improprieties of language, and, if one never dared to raise oneself above abuses of this kind, one could never even try to introduce more exactitude and precision to terms than that which they carry in current usage.

and 'unassignable' are truly and completely synonymous; but in any case, it is at the very least certain that a truly 'unassignable' quantity, by reason of the possibility of indefinite decrease that it implies, will thereby be 'incomparable' with respect to any given quantity, and, to extend the idea to different orders of the infinitesimal, even with respect to any quantity in relation to which it can decrease indefinitely, as long as the latter is regarded as possessing at least a relative fixity.

If there is one point on which everyone can easily agree, even without going more deeply into questions of principles, it is that the notion of the indefinitely small, at least from the purely mathematical point of view, is perfectly sufficient for infinitesimal analysis, and the 'infinitists' themselves recognize this without great difficulty.[7] In this respect one can thus be content with a definition such as that given by Carnot: 'What is an infinitely small quantity in mathematics? Nothing other than a quantity that can be rendered as small as one likes, without one's being obliged on that account to vary those to which one wants to relate it.'[8] But as for the true significance of infinitesimal quantities, the entire matter is not limited to this; for the calculus it matters little that the infinitely small are only fictions, since one can be content with a consideration of the indefinitely small, which raises no logical difficulty; furthermore, since for the metaphysical reasons set out at the beginning we cannot admit a quantitative infinite, whether infinitely great or infinitely small,[9] or indeed any infinite of a determined and relative order, it is quite certain that these can only be fictions and nothing else; but if rightly

7. See especially L. Couturat, *De l'infini mathématique*, p265, note: 'One can logically constitute the infinitesimal calculus on the sole notion of the indefinite. . . .' It is true that the use of the word 'logically' here implies a reservation, for it is opposed, for the author, to 'rationally', which is moreover a rather strange terminology; the admission is nonetheless interesting to keep in mind.

8. *Réflexions sur la Métaphysique du Calcul infinitésimal*, p7, note; cf. ibid., p20. The title of this work is scarcely justified, for in reality there is not to be found in it the least idea of a metaphysical order.

9. Pascal's overly celebrated conception of 'two infinities' is metaphysically absurd, and it is again only the result of a confusion of the infinite with the indefinite, the latter being taken in the two opposite directions of increasing and decreasing magnitude.

39

or wrongly these fictions were introduced into the infinitesimal calculus in the beginning, this is because according to Leibniz's intention they nevertheless correspond to something, however faulty the manner in which they expressed it. Since we are here concerned with principles and not merely with a method of calculation in and of itself (which would be without interest for us) we should therefore ask what exactly is the value of these fictions, not only from the logical point of view, but also from the ontological point of view, whether they are as 'well-founded' as Leibniz believed, and whether we can even say with him that they are *toleranter verae*, and at the very least accept them as such *modo sano sensu intelligantur* [understood in a reasonable way]. To answer these questions it will be necessary for us to examine more closely his conception of the 'law of continuity', for it was here that he thought to find the *fundamentum in re* of the infinitely small.

7

Degrees of Infinity

We have not yet had occasion in the preceding pages to see all the confusions that are inevitably introduced when the idea of the infinite is taken otherwise than in its one true and properly metaphysical sense; more than one example could be found, notably, in Leibniz's long discussion with Jean Bernoulli on the reality of infinite and infinitely small quantities, which moreover never came to any definitive conclusion; nor, indeed, could it have done so, given the continual confusion on both sides, and the lack of principles from which this confusion proceeded; moreover, whatever the order of ideas in question, ultimately it is always the lack of principles which alone renders questions insoluble. One might well be astonished to learn, among other things, that Leibniz distinguished between 'infinite' and 'interminable', and that he had thus not absolutely rejected the idea—nonetheless manifestly contra-dictory—of a 'terminating infinite', and went so far as to ask himself 'whether it might be possible for there to exist, for example, an infinite straight line that might nevertheless terminate at both ends.'[1] No doubt he is reluctant to admit this possibility, 'all the more so since it seems to me,' he says elsewhere, 'that the infinite, taken rigorously, must have its source in the interminable, without which I see no means of finding a proper foundation for distinguishing it from the finite.'[2] But even if one puts it more affirmatively (which he did not do) and says that 'the infinite has its source in the interminable,' one still does not take them to be absolutely identical, but rather as distinguished from one another to a certain degree; and as long as that is

1. Letter to Jean Bernoulli, November 18, 1698.
2. Previously cited letter to Varignon, February 2, 1702.

41

so, one risks finding oneself checked by a crowd of strange and contradictory ideas. It is true that Leibniz declares that he would not willingly admit these ideas without first being 'forced by indubitable demonstrations', but it is already serious enough to attribute a certain degree of importance to them, and even to be able to envisage them other than as pure impossibilities; as for the idea of a sort of 'terminating eternity', to take one example from those he sets forth in this connection, we can see in it only the product of a confusion between the notions of eternity and duration, which is absolutely unjustifiable with respect to metaphysics. We readily grant that the time in which we pass our corporeal lives really is indefinite, which is in no way incompatible with its 'terminating at both ends', which is to say, in conformity with the traditional cyclic conception, that it has both a beginning and an end; we also grant that there exist other modes of duration, such as that which the Scholastics call *aevum*, the indefinitude of which is, if one may so express it, indefinitely greater than that of this time; but all these modes, in all their possible extension, are nonetheless only indefinite, since it is always a question of particular conditions of existence proper to this or that state; and, precisely insofar as each is a kind of duration—which implies succession—not one can be identified with or assimilated to eternity, with which it has no more connection than does the finite, whatever its mode, nor again with the true Infinite, for the notion of a relative eternity has no more meaning than that of a relative infinite. In all of this we have only various orders of indefinitude, as will be seen more clearly later on, but Leibniz, for want of having made the necessary and essential distinctions, and above all for not having laid down before all else the principle that alone would have prevented him from going astray, found himself very much at a loss to refute Bernoulli's opinions; indeed, so equivocal and hesitant were Leibniz's responses that Bernoulli even took him to be much closer than was really the case to his own ideas about the 'infinity of worlds' and the different 'degrees of infinity'.

This notion of the so-called 'degrees of infinity' amounts in short to supposing that there can exist worlds incomparably greater and incomparably smaller than our own, the corresponding parts of each being in equal proportion to one another, such that the inhabitants of any one of these worlds would have just as much reason to call theirs infinite as we would with respect to ours; for our part we would rather say they would have just as little reason. Such a manner of envisaging things would not appear absurd *a priori* without the introduction of the idea of the infinite, which is certainly nothing to the purpose, for however great one imagines them to be, each of these worlds is nonetheless limited; how then can they be called infinite? The truth is that none of them can really be so, if only because they are conceived as multiple, for here we return to the contradiction of a plurality of infinites; and besides, even if it happens that some or even many consider our world to be infinite, this assertion nonetheless can offer no acceptable meaning. Moreover, one might wonder if they really are different worlds, or if, quite simply, they are not rather more or less extended parts of the same world, since by hypothesis they must all be subject to the same conditions of existence—notably to spatiality—and simply developed on an enlarged or diminished scale. It is in a completely different sense that one can truly speak, not of an infinity, but of an indefinitude of worlds, since apart from the conditions of existence such as space and time, which are proper to our world considered in all the extension of which it is susceptible, there is an indefinitude of others, equally possible; a world, or, in short, a state of existence, is thus defined by the totality of the conditions to which it is subject; but, by the very fact that it will always be conditioned, that is, determined and limited, and hence unable to contain all possibilities, it can never be regarded as infinite, but only indefinite.[3]

Fundamentally, the consideration of 'worlds' in the sense understood by Bernoulli, incomparably larger or smaller in relation to one another, is not very different from what Leibniz resorted to when he envisaged 'the firmament with respect to the earth, and the earth with respect to a grain of sand,' and the latter with respect to 'a particle of magnetic material passing through a lens.' Only here Leibniz does not claim to speak of *gradus infinitatis* [grade of infinity] in the strict sense; on the contrary, he even means to show that 'one need not take the infinite rigorously,' and he is content to

3. On this subject, see *The Multiple States of the Being*.

envisage 'incomparables', to which no logical objection can be raised. The shortcoming of his comparison is of quite another order, and as we have already said, lies in the fact that it is only capable of giving an inexact, or even completely false, idea of the infinitesimal quantities as they figure in the calculus. In what follows we shall have occasion to substitute for this consideration that of the true multiple degrees of indefinitude, taken in increasing as well as decreasing order; for the moment, therefore, we shall not dwell further on it.

In short, the difference between Bernoulli and Leibniz is that for the first, even though he presents them only as a probable conjecture, it is truly a question of 'degrees of infinity', while the second, doubting their probability and even their possibility, limits himself to replacing them with what could be called 'degrees of incomparability'. Aside from this difference, which is moreover assuredly extremely important, they share in common the notion of a series of worlds that are similar, but on different scales. This notion is not without a certain incidental connection with discoveries made in the same period with the microscope, and with certain views that arose as a consequence—although later observations were in no way to justify them—such as the theory of the 'encasement of embryos'; now it is not true of embryos that every part of the living being must be actually and physically 'preformed', and the organization of a cell bears no resemblance to that of the entire body of which it is an element. There seems to be no doubt that this was in fact the origin of Bernoulli's notion, at any rate; indeed, among other things highly significant in this regard, he says that the particles of a body coexist in the whole 'in the same way that, in accordance with Harvey and others, though not with Leeuwenhoeck, there exist within an animal innumerable ovules, within each ovule one or several animalcules, within each animalcule again innumerable ovules, and so on to infinity.'[4] As for Leibniz, his was likely a completely different point of departure; thus, the idea that all the stars that we can see can only be components of the body of an incomparably greater being, recalls the Kabbalistic conception of the 'Great Man', but singularly

4. Letter of July 23, 1698.

44

materialized and 'spatialized' through a sort of ignorance of the true analogical value of traditional symbolism; likewise, the idea of the 'animal', that is, the living being, subsisting corporeally after death, but 'in miniature', is obviously inspired by the traditional Judaic concept of the *luz* or 'kernel of immortality',[5] which Leibniz equally distorted by connecting it with the notion of worlds incomparably smaller than our own, saying, 'nothing prevents animals from being transferred to such worlds after death; indeed, I think that death is no more than a contraction of the animal, just as generation is no more than an evolution,'[6] this last word being taken here simply in its etymological sense of 'development'. All this is fundamentally only an example of the dangers that exist when one wishes to make traditional notions agree with the views of profane science, which can only be done to the detriment of the former; these notions are most clearly independent of the theories brought about by microscopic observations, and in comparing and muddling them, Leibniz was already acting as would the occultists later on, for they particularly delighted in these sorts of unjustified comparisons. Moreover, the superposition of 'incomparables' of different orders seemed to him in conformity with his notion of the 'best of worlds', furnishing a means of investing it with 'as much being or reality as possible', to quote from his definition; and as we have already pointed out elsewhere,[7] this idea of the 'best of worlds' is also derived from yet another ill-applied traditional doctrine, this one borrowed from the symbolic geometry of the Pythagoreans. According to this geometry, of all lines of equal length, the circumference of a circle is that which encloses the maximum surface area, and of all bodies of equal surface area, the sphere is likewise that which contains the maximum volume, and this is one of the reasons why these figures were regarded as the most perfect. But if in this respect there is a maximum, there is nonetheless no minimum, that is, there exist no fig-

5. See *The King of the World*, chap. 7.
6. Previously cited letter to Jean Bernoulli, November 18, 1698.
7. *The Symbolism of the Cross*, chap. 6. On the distinction between 'possibles' and 'compossibles', on which the notion of the 'best of worlds' further depends, cf. *The Multiple States of the Being*, chap. 2.

ures enclosing a surface area or a volume less than all others, and this is why Leibniz was led to think that, although there is a 'best of worlds', there is no 'worst of worlds', that is, a world containing less being than any other possible world. Moreover, we know that this notion of the 'best of worlds', like that of 'incomparables', is linked to the well-known comparisons involving the 'garden full of plants' and the 'pond filled with fish', where 'each twig of the plant, each member of the animal, each drop of its humors, is again such a garden or such a pond';[8] and this naturally brings us to another, related question, that of the 'infinite division of matter'.

8. *Monadologie*, 67; cf. ibid., 74.

8

Infinite Division
or Indefinite Divisibility

For Leibniz, not only is matter divisible, but all its parts are 'actually sub-divided without end, . . . each part into parts, each having some movement of its own';[1] and he emphasizes this point above all in order to offer theoretical support to the concept we last explained: 'It follows from the actual division that in every part of matter, however small it might be, there is as it were a world consisting of innumerable creatures.'[2] Bernoulli likewise supposes this actual division of matter *in partes numero infinitas* [into infinitely many parts], but he draws from it conclusions Leibniz did not accept: 'If a finite body,' he says, 'has parts infinite in number, I have always believed, and still do, that the smallest of these parts must have an unassignable, or infinitely small, ratio to the whole';[3] to which Leibniz responds: 'Even if one agrees that there is no portion of matter that is not actually divided, one does not, however, arrive at indivisible elements, or at parts smaller than all others, or infinitely small, but only at ever smaller parts, which, however, are ordinary quantities, just as in augmentation one arrives at ever greater quantities.'[4] Thus it is the existence of *minimae portiones* [smallest parts], or of 'final elements', that Leibniz contests; for Bernoulli, on the contrary, it seems clear that actual division implies the simultaneous existence of all the elements in question, just as, if an 'infinite' sequence

1. *Monadologie*, 65.
2. Letter to Jean Bernoulli, July 12–22, 1698.
3. Previously cited letter of July 23, 1698.
4. Letter of July 29, 1698.

47

be given, all of its constituent terms must be given simultaneously, which implies the existence of a *terminus infinitesimus* [infinitesimal limit]. But for Leibniz the existence of this limit is no less contradictory than that of an 'infinite number', and the notion of a smallest of numbers, or a *fractio omnium infima* [a part smaller than all others], no less absurd than that of a greatest of numbers. What he considers to be the 'infinity' of a sequence is characterized by the impossibility of arriving at a final term, and matter would likewise not be 'infinitely' divided if this division could ever be completed and end at 'final elements'; and it is not only that we could not in fact ever arrive at these final elements, as Bernoulli concedes, but that they should not exist in nature at all. There are no indivisible corporeal elements, or 'atoms' in the proper sense of the word, any more than there are indivisible fractions that cannot yield ever smaller fractions in the numerical order, or, in the geometric order, linear elements that cannot be divided into ever smaller elements.

In all of this Leibniz basically takes the word 'infinite' in exactly the same sense as he does when speaking of an 'infinite multitude'; for him, to say of any sequence, including that of the whole numbers, that it is infinite is not to say that it must come to a *terminus infinitesimus* or an 'infinite number', but on the contrary that it must have no final term, since its terms are *plus quam numero designari possint* [more than can be numbered], that is, they constitute a multitude that surpasses all number. Similarly, if one can say that matter is infinitely divided, this is because any one of its portions, however small, always encloses such a multitude; in other words, matter does not have *partes minimae* [smallest parts] or simple elements, it is essentially a composite: 'It is true that simple substances, that is, those that do not exist by aggregation, really are indivisible, but they are immaterial, and are only principles of action.'[5] It is in the sense of an innumerable multitude—which, moreover, is the sense Leibniz most commonly employs—that the idea of the so-called infinite can be applied to matter, to geometric extension, and in general to the continuous, taken in relation to its composition; besides, this sense is not exclusive to the *infinitum*

5. Letter to Varignon, June 20, 1702.

continuum [continuous infinite] but extends to the *infinitum discretum* [discrete infinite] as well, as we have seen both in the example of the multitude of all the numbers and in that of the 'infinite sequence'. This is why Leibniz was able to say that a magnitude is infinite insofar as it is 'inexhaustible', which means that 'one can always take a magnitude as small as one likes', and, 'it remains true, for example, that 2 is as much as $1/1 + 1/2 + 1/4 + 1/8 + 1/16 + 1/32 + \ldots$, which is an infinite series, comprised at once of all fractions with a numerator of 1 and denominators in double geometric progression, although only ordinary numbers are ever used, that is, one never introduces any infinitely small fraction, or one with an infinite number for its denominator.'[6] Moreover, what was just said allows us to understand how Leibniz, while affirming that the infinite, as he understands it, is not a whole, nevertheless could apply this idea to the continuous: a continuous set, as any given body, indeed constitutes a whole, even what we above called a true whole, logically anterior to its parts and independent of them, but it is obviously always finite as such; it is therefore not with respect to the whole that Leibniz is able to call it infinite, but only with respect to its parts into which it can be divided, and only insofar as the multitude of these parts effectively surpasses every assignable number. This is what one might call an analytical conception of the infinite, since in fact, it is only analytically that the multitude in question is inexhaustible, as we shall explain later.

If we now question the worth of the idea of 'infinite division', we must recognize that, as with the 'infinite multitude', it contains a certain portion of truth, though its manner of expression is anything but safe from criticism. First of all, it goes without saying that, in accordance with all that we have explained so far, there can be no question of infinite division, but only of indefinite division; and on the other hand it is necessary to apply this idea not to matter in general, which would perhaps have no meaning, but only to bodies, or to corporeal matter if one insists on speaking of 'matter' here, in spite of the extreme obscurity of the notion, and the many equivocations to which it gives rise.[7] In fact, it is to extension that divisibil-

---

6. Previously cited letter to Varignon, February 2, 1702.

ity properly pertains, not to matter, in whatever sense this is understood, and the two could only be confused were one to adopt the Cartesian concept, according to which the nature of bodies consists essentially and uniquely in extension, a concept, moreover, that Leibniz also did not admit. If, then, all bodies are necessarily divisible, this is because they possess extension, and not because they are material. Now let us again recall that extension, being something determined, cannot be infinite; hence, it obviously cannot imply any possibility more infinite than itself; but as divisibility is a quality inherent to the nature of extension, its limitations can only come from this nature itself; as long as there is extension, it is always divisible, and one can thus consider its divisibility to be truly indefinite, its indefinitude being conditioned, moreover, by that of extension. Consequently, extension as such cannot be composed of indivisible elements, for these elements would have to be extensionless to be truly indivisible, and a sum of elements with no extension can no more constitute an extension than a sum of zeros can constitute a number; this is why, as we have explained elsewhere,[8] points are not the elements or parts of a line; the true linear elements are always distances between points, which latter are only their extremities. Moreover, Leibniz himself envisaged things thus in this regard, and according to him, this is precisely what marks the fundamental difference between his infinitesimal method and Cavalieri's 'method of indivisibles', namely, that he does not consider a line to be composed of points, or a surface of lines, or a volume of surfaces: points, lines, and surfaces are here only limits or extremities, not constituent elements. It is indeed obvious that points, multiplied by any quantity at all, can never produce length, since, rigorously speaking, they are null with respect to length; the true elements of a magnitude must always be of the same nature as the magnitude, although incomparably less: this leaves no room for 'indivisibles', and what is more, it allows us to observe in the infinitesimal calculus a certain law of homogeneity, which implies that ordinary quantities and infinitesimal quantities of various orders, although

7. On this subject, see *The Reign of Quantity and the Signs of the Times*.
8. *The Symbolism of the Cross*, chap. 16.

incomparable among themselves, are nonetheless magnitudes of the same species.

From this point of view one can say in addition that the part, whatever it be, must always preserve a certain 'homogeneity' or conformity of nature with the whole, at least insofar as the whole is considered able to be reconstituted by means of its parts, by a procedure comparable to that used in the formation of an arithmetical sum. Moreover, this is not to say that no simple thing exists in reality, for composites can be formed, starting from their elements, in a way completely different from this; but then, to speak truly, these elements are no longer properly 'parts', and as Leibniz recognized, they can in no way be of a corporeal order. What is indeed certain is that one cannot arrive at simple, that is, indivisible, elements without departing from the special condition that is extension; the latter could not be resolved into such elements without ceasing to be as extension. It immediately follows that there cannot exist indivisible corporeal elements, as this notion implies a contradiction; for indeed, such elements would have to be without extension, and then they would no longer be corporeal, for by very definition the word 'corporeal' necessarily entails extension, although this is not the whole nature of bodies; thus, despite all the reservations we must make in other regards, Leibniz is at least entirely right in his position against atomism.

But until now we have spoken only of divisibility, that is to say the possibility of division; must we go further and admit with Leibniz an 'actual division'? This idea is also not exempt from contradiction, for it amounts to supposing an entirely realized indefinite and on that account is contrary to the very nature of indefinitude, which, as we have said, is always a possibility in the process of development, hence essentially implying something unfinished, not yet completely realized. Moreover, there is in fact no reason to make such a supposition, for when presented with a continuous set we are given the whole, not the parts into which it can be divided, and it is only we who conceive that it is possible for us to divide this whole into parts capable of being rendered smaller and smaller so as to become less than any given magnitude, provided the division be carried far enough; in fact, it is consequently we who realize the parts, to the extent that we effectuate the division. Thus, what exempts us from having to suppose an 'actual division' is the distinction we established earlier on the subject of the different ways of envisaging a whole: a continuous set is not the result of the parts into which it is divisible but is on the contrary independent of them, and, consequently, the fact that it is given to us as a whole by no means implies the actual existence of those parts.

Likewise, from another point of view and passing on to a consideration of the discontinuous, we can say that if an indefinite numerical sequence is given, this in no way implies that all the terms it contains are given distinctly, which is impossible precisely inasmuch as it is indefinite; in reality, to give such a sequence is simply to give the law that enables one to calculate the term occupying a determined position, or, for that matter, any position whatsoever within the sequence.[9] If Leibniz had given this answer to Bernoulli, their discussion on the existence of the *terminus infinitesimus* would thereby have been brought to an immediate close; but he would not have been able to do so without logically being led to renounce his idea of 'actual division', unless he were to deny all correlation between continuous and discontinuous modes of quantity.

Be that as it may, as far as the continuous is concerned at any rate, it is precisely in the 'indistinction' of its parts that we can see the root of the idea of the infinite such as it was understood by Leibniz, since, as we said earlier, this idea always carries with it a certain amount of confusion; but this 'indistinction', far from presupposing

9. Cf. L. Couturat, *De l'infini mathématique*, p467: 'The sequence of natural numbers is given entirely by its law of formation, as moreover is the case with all other infinite sequences and series: in general a formula of recurrence suffices to define them entirely, such that their limit or sum (when it exists) is on that account completely determined. . . . It is thanks to this law of formation of the sequence of natural numbers that we have an idea of every whole number, and in this sense they are altogether given by this law.'—One can indeed say that the general formula expressing the nth term of a sequence contains, potentially and implicitly, though not actually and distinctly, all the terms of the sequence, since any of them can be derived from it by giving to n the value corresponding to the position the term occupies in the sequence; but, contrary to what Couturat thought, this is certainly not what Leibniz meant to say 'when he maintained the actual infinity of the sequence of natural numbers.'

a realized division, tends on the contrary to exclude it, even apart from the completely decisive reasons we have just noted. Therefore, even if Leibniz's theory is right insofar as it is opposed to atomism, it must be corrected elsewhere if it is to correspond to truth; the 'infinite division of matter' must be replaced by the 'indefinite divisibility of extension'; here, in its briefest and most precise expression, is the conclusion to which all the considerations we have just set forth ultimately lead.

9

Indefinitely Increasing; Indefinitely Decreasing

BEFORE CONTINUING the examination of questions properly relating to the continuous, we must return to what was said above about the non-existence of a *fractio omnium infima*, which will allow us to see how the correlation or symmetry that exists in certain respects between indefinitely increasing and indefinitely decreasing quantities can be represented numerically. We have seen that in the domain of discontinuous quantity, as long as it is only the sequence of whole numbers that needs to be considered, these numbers must be regarded as increasing indefinitely starting from the unit, but that there can obviously be no question of an indefinite decrease since the unit is essentially indivisible; were the numbers to be taken in the decreasing direction, one would necessarily find oneself stopped at the unit itself, so that the representation of the indefinite by whole numbers is limited to a single direction, that of indefinite increase. On the other hand, when it is a question of continuous quantity, one can envisage indefinitely decreasing quantities as well as indefinitely increasing ones; and the same occurs in discontinuous quantity itself as soon as, in order to express this possibility, the consideration of fractional numbers is introduced. Indeed, one can envisage a sequence of fractions continuing to decrease indefinitely; that is, however small a fraction might be, a smaller one could always be formed, and this decrease can no more arrive at a *fractio minima* [smallest fraction] than can the increase of whole numbers at a *numerus maximus* [greatest number].

If we wish to use a numerical representation in order to make evident the correlation between the indefinitely increasing and the

54

indefinitely decreasing, it suffices to consider the sequence of whole numbers together with that of their inverses; a number is said to be the inverse of another when the product of the two is equal to the unit, and for this reason the inverse of the number n is represented by the notation $1/n$. Whereas the sequence of whole numbers goes on increasing indefinitely starting from the unit, the sequence of their inverses decreases indefinitely, starting from the same unit, which is its own inverse, and which is therefore the common point of departure for the two sequences; to each number in one sequence there thus corresponds a number in the other, and inversely, such that the two sequences are equally indefinite, and in exactly the same way, though in contrary directions. The inverse of a number is obviously as small as the number itself is great, since their product always remains constant; however great a number n might be, the number $n+1$ will be greater still by virtue of the very law of formation for the indefinite sequence of whole numbers, and similarly, as small as a number $1/n$ might be, the number $1/(n+1)$ will be smaller still; and this clearly proves the impossibility of any 'smallest of numbers', which notion is no less contradictory than is that of a 'greatest of numbers', for, if it is impossible to stop at a determined number in the increasing direction, it will be no more possible to stop in the decreasing direction. Moreover, since this correlation which is found in numerical discontinuity occurs first of all as a consequence of the application of this discontinuity to the continuous, as we said concerning fractional numbers, the introduction of which it naturally supposes, it can only express the correlation that exists within the continuous itself between the indefinitely increasing and the indefinitely decreasing in its own way, which is necessarily conditioned by the nature of number. Therefore, whenever continuous quantities are considered capable of becoming as great or as small as one likes, that is, greater or smaller than any determined quantity, one can always observe a symmetry and, in a manner of speaking, a parallelism presented by these two inverse kinds of variability. This remark will subsequently help us to understand better the possibility of different orders of infinitesimal quantities.

It would be good to point out that although the symbol $1/n$

evokes the idea of fractional numbers, and although it is in fact incontestably derived from them, the inverses of the whole numbers need not be defined here as such, and this in order to avoid the difficulty presented by the ordinary notion of fractional numbers from the strictly arithmetical point of view, that is, the conception of fractions as 'parts of the unit'. Indeed, it suffices to consider the two sequences to be constituted by numbers respectively greater and smaller than the unit, that is, as two orders of magnitude that have their common limit in the latter, and that at the same time both can be regarded as issuing from this unit, which is truly the primary source of all numbers; what is more, if one wished to consider the two indefinite sets as forming a single sequence, one could say that the unit occupies the exact mid-point within this sequence, since, as we have seen, there are exactly as many numbers in the one set as in the other. Moreover, if, to generalize further, instead of considering only the sequence of whole numbers and their inverses, one wished to introduce fractional numbers properly speaking, nothing would be changed as far as the symmetry of increasing and decreasing quantities is concerned: on one side one would have all the numbers greater than the unit, and on the other all those smaller than the unit; here, again, for any number $a/b > 1$, there will be a corresponding number $b/a < 1$ in the other group, and reciprocally, such that $(a/b)\,(b/a) = 1$, just as earlier we had $(n)\,(1/n) = 1$, and there will thus be exactly the same number of terms in each of these two indefinite groups separated by the unit; it must moreover be understood that when we say 'the same number of terms', we simply mean that the two multitudes correspond term by term, and not that they can themselves on that account be considered 'numerable'. Any two inverse numbers multiplied together always produce again the unit from which they proceeded; one can say further that the unit, occupying the mid-point between the two groups, and being the only number that can be regarded as belonging to both at once[1]—

1. According to the definition of inverse numbers, the unit appears first in the form 1 and then again in the form ⅟₁, such that $(1)\,(1/1) = 1$; but, as on the other hand $1/1 = 1$, it is the same unit that is thus represented in two different forms, and it is consequently, as we said above, its own inverse.

although in reality it would be more correct to say that it unites rather than separates them—corresponds to the state of perfect equilibrium, and contains in itself all numbers which issue from it in pairs of inverse or complementary numbers, each pair by virtue of this complementarity constituting a relative unity in its indivisible duality;[2] but we shall return a little later to this last consideration and to the consequences it implies.

Instead of saying that the series of whole numbers is indefinitely increasing and that of their inverses indefinitely decreasing, one could also say, in the same sense, that the numbers thus tend on the one hand toward the indefinitely great and on the other toward the indefinitely small, on condition that we understand by this the actual limits of the domain in which these numbers are considered, for a variable quantity can only tend toward a limit. The domain in question is, in short, that of numerical quantity, taken in every possible extension;[3] this again amounts to saying that its limits are not determined by such and such a particular number, however great or small it might be supposed, but by the very nature of number as such. By the very fact that number, like everything else of a determined nature, excludes all that it is not, there can be no question of the infinite; moreover, we have just said that the indefinitely great must inevitably be conceived of as a limit, although it is in no way a *terminus ultimus* [ultimate limit] of the series of numbers, and in this connection one can point out that the expression 'tend toward infinity', frequently employed by mathematicians in the sense of 'increase indefinitely', is again an absurdity, since the infinite obviously implies the absence of any limit, and that consequently there is nothing toward which it is possible to tend. What is also rather remarkable is that certain mathematicians, while recognizing the

2. We say indivisible because whenever one of the two numbers forming such a pair exists, the other also necessarily exists by that very fact.

3. It goes without saying that the incommensurable numbers, in relation to magnitude, are necessarily interspersed among the ordinary numbers, which are whole or fractional according to whether they are greater or smaller than the unit; this demonstrates, moreover, the geometrical correspondence we pointed out earlier, as well as the possibility of defining such a number by two convergent sets of commensurable numbers, of which it is the common limit.

inaccuracy and improper character of the expression 'tend toward infinity', on the other hand feel no scruple at all about taking the expression 'tend toward zero' in the sense of 'decrease indefinitely'; zero, however, or the 'null quantity', is, with respect to decreasing quantities, exactly the same as the so-called 'quantitative infinite' is with respect to increasing quantities; but we shall have to return to these questions later, particularly when we come to the subject of zero and its different meanings.

Since the sequence of numbers in its entirety is not 'terminated' by a given number, it follows that there is no number however great that could be identified with the indefinitely great in the sense just understood; and, naturally, the same is true for the indefinitely small. One can only regard a number as practically indefinite, if one may so express it, when it can no longer be expressed by language or represented by writing, which in fact inevitably occurs the moment one considers numbers that go on increasing or decreasing; here we have a simple matter of 'perspective', if one wishes, but all in all even this is in keeping with the character of the indefinite, insofar as the latter is ultimately nothing other than that of which the limits can be, not done away with, since this would be contrary to the very nature of things, but simply pushed back to the point of being entirely lost from view. In this connection some rather curious questions should be considered; thus, one could ask why the Chinese language symbolically represents the indefinite by the number ten thousand; the expression 'the ten thousand beings', for example, means all beings, which really make up an indefinite or 'innumerable' multitude. What is quite remarkable is that it is precisely the same in Greek, where a single word likewise serves to express both ideas at once, with a simple difference in accentuation, obviously only a quite secondary detail, and doubtless only due to the need to distinguish the two meanings in usage: μύριοι, 'ten thousand'; μυρίοι, 'an indefinitude'. The true reason for this is the following: the number ten thousand is the fourth power of ten; now, according to the formulation of the *Tao Te Ching*, 'one produced two, two produced three, three produced all numbers,' which implies that four, produced immediately after three, is in a way equivalent to the whole set of numbers, and this because, when one has the quaternary by adding the first four numbers, one also has the denary, which represents a complete numerical cycle: $1+2+3+4=10$, which, as we have already said on other occasions, is the numerical formula of the Pythagorean *Tetraktys*. One can further add that this representation of numerical indefinitude has its correspondence in the spatial order: it is common knowledge that raising a number from one degree to the next highest power represents in this order the addition of a dimension; now, our space having only three dimensions, its limits are transcended when one goes beyond the third power, which, in other words, amounts to saying that elevation to the fourth power marks the very term of its indefinitude, since, as soon as it is effected, one has thereby departed from space and passed on to another order of possibilities.

10

The Infinite
and the Continuous

The idea of the infinite as Leibniz most often understood it, which, let us never forget, was merely that of a multitude surpassing all number, sometimes appears under the aspect of a 'discontinuous infinite', as in the case of so-called infinite numerical sequences; but its most usual aspect, and also its most important one as far as the significance of the infinitesimal calculus is concerned, is that of the 'continuous infinite'. In this regard it is useful to recall that when Leibniz, beginning the research that at least according to what he himself said, would lead to the discovery of his method, was working with sequences of numbers, he at first considered only differences that are 'finite' in the ordinary sense of the word; infinitesimal differences appeared to him only when there was a question of applying numerical discontinuity to the spatial continuum. The introduction of differentials was therefore justified by the observation of a certain analogy between the respective kinds of variability within these two modes of quantity; but their infinitesimal character arose from the continuity of the magnitudes to which they had to be applied, and thus, for Leibniz, a consideration of the 'infinitely small' is closely linked to that of the 'composition of the continuous'.

Taken 'rigorously', 'infinitely small', would be *partes minimae* of the continuous, as Bernoulli thought; but clearly the continuous, insofar as it exists as such, is always divisible, and consequently it could not have *partes minimae*. 'Indivisibles' cannot even be said to be parts of that with respect to which they are indivisible, and 'minimum' can be understood here only as a limit or extremity, not as an element: 'Not only is a line less than any surface,' Leibniz says, 'it is not even part of a surface, but merely a minimum or an extremity';[1] and from his point of view this assimilation between *extremum* and *minimum* can be justified by the 'law of continuity', in that according to him it permits 'passage to the limit', as we shall see later. As we have said already, the same holds for a point with respect to a line, as well as for a surface with respect to a volume; on the other hand, the infinitesimal elements must be parts of the continuous, without which they could not even be quantities; and they can be so only on condition of not truly being 'infinitely small', for then they would be nothing other than *partes minimae* [smallest parts] or 'final elements', of which the very existence implies a contradiction in regard to the continuous. Thus the composition of the continuous prevents infinitely small quantities from being anything more than simple fictions; but from another point of view, it is nevertheless precisely the existence of this continuity that makes them 'well-founded fictions', at least in Leibniz's eyes: if 'within the realm of geometry they may be treated as if they were perfectly real,' this is because extension, which is the object of geometry, is continuous; and, if it is the same with nature, this is because bodies are likewise continuous, and also because there is also continuity in all phenomena such as movement, of which these bodies are the seat, and which are the objects of mechanics and physics. Moreover, if bodies are continuous, this is because they are extended and participate in the nature of extension; and similarly, the continuity of movement, as well as of the various phenomena more or less directly connected to it, derives essentially from its spatial character. Thus the continuity of extension is ultimately the true foundation of all other continuity that is observed in corporeal nature; and this, moreover, is why in introducing an essential distinction that Leibniz did not make in this regard, we specified that in reality one must attribute

---

1. *Meditatio nova de natura anguli contactus et osculi, horumque usu in practica Mathesi ad figuras faciliores succedaneas difficilioribus substituendas* [A New Reflection on the Nature of Angles of Contact and Tangency and on the Use of These in Practical Mathematics for Substituting Easier Figures for the More Difficult], in the *Acta Eruditorum* of Leipzig, 1686.

the property of 'indefinite divisibility' not to 'matter' as such, but rather to extension.

Here we need not examine the question of other possible forms of continuity, independent of its spatial form; indeed, one must always return to the latter when considering magnitudes, and its consideration thus suffices for all that pertains to infinitesimal quantities. We should, however, include together with it the continuity of time, for contrary to the strange opinion of Descartes on the subject, time really is continuous in and of itself, and not merely with respect to its spatial representation in the movement used to measure it.[2] In this regard, one could say that movement is as it were doubly continuous, for it is so in virtue both of its spatial and of its temporal condition; and this sort of combination of space and time, from which movement results, would not be possible were the one discontinuous and the other continuous. This consideration also allows the introduction of continuity into various categories of natural phenomena that pertain more directly to time than to space, although occurring in both, as, for example, with any processes of organic development. As for the composition of the temporal continuum, moreover, one could repeat everything said concerning the composition of the spatial continuum, and in virtue of this sort of symmetry which, as we have seen, exists in certain respects between space and time, one will arrive at strictly analogous conclusions; instants conceived of as indivisible are no more parts of duration than are points of extension, as Leibniz likewise recognized, and here again we have a thesis with which the Scholastics were quite familiar; in short, it is a general characteristic of all continuity that its nature precludes the existence of 'final elements'.

All that we have said up to this point sufficiently shows in what sense one may understand that from Leibniz's point of view, the continuous necessarily embraces the infinite; but we cannot, of course, suppose that there is any question of an 'actual infinity', as if all possible parts are effectively given whenever a whole is given; nor is there any question of a true infinity, which any determination whatsoever would exclude, and which consequently cannot be

2. Cf. *The Reign of Quantity and the Signs of the Times*, chap. 5.

implied by the consideration of any particular thing. Here, however, as in every case in which the idea of an alleged infinite presents itself, different from the true metaphysical Infinite, but in itself representing something other than a pure and simple absurdity, all contradiction disappears, and with it all logical difficulty, if one replaces the so-called infinite with the indefinite, and if one simply says that all continuity, when taken with respect to its elements, embraces a certain indefinitude. It is also for lack of having made this fundamental distinction between the Infinite and the indefinite that some people have mistakenly believed it impossible to escape the contradiction of a determined infinite except by rejecting the continuous altogether and replacing it with the discontinuous; thus Renouvier, who rightly denied the mathematical infinite, but to whom the idea of the metaphysical Infinite was nevertheless completely foreign, believed that the logic of his 'finitism' obliged him to go so far as to accept atomism, thus falling prey to a concept no less contradictory than the one he wished to avoid, as we saw earlier.

11

The
Law of Continuity

Whenever there exists a continuum, we can say with Leibniz that there is something of the continuous in its nature, or, if one prefer, that there must be a certain 'law of continuity' applying to all that presents the characteristics of the continuous; this is obvious enough, but it by no means follows that such a law must then be applicable to absolutely everything, as he claims, for, if the continuous exists, so does the discontinuous, even in the domain of quantity;[1] number, indeed, is essentially discontinuous, and it is this very discontinuous quantity, and not continuous quantity, that is really the first and fundamental mode of quantity, what one might properly call pure quantity, as we have said elsewhere.[2] Moreover, nothing allows us to suppose *a priori* that, outside of pure quantity, a continuity of some kind exists everywhere, and, to tell the truth, it would be quite astonishing if, among all possible things, number alone had the property of being essentially discontinuous; but our

1. Cf. L. Couturat, *De l'infini mathématique*, p140: 'In general, the principle of continuity has no place in algebra, and cannot be invoked to justify the algebraic generalization of number. Not only is continuity by no means necessary to speculations concerning general arithmetic, it is repugnant to the spirit of the science, and to the very nature of number. Number, indeed, is essentially discontinuous, as are nearly all its arithmetical properties.... One therefore cannot impose continuity on algebraic functions, however complicated they might be, since the whole numbers, which furnish their elements, are discontinuous, "jumping", as it were, from one value to the next without any possible transition.'
2. See *The Reign of Quantity and the Signs of the Times*, chap. 11.

intention is not to determine the bounds within which a 'law of continuity' truly is applicable, or what restrictions should be brought to bear on all that goes beyond the domain of quantity understood in its most general sense. We shall limit ourselves to giving one very simple example of discontinuity, taken from the realm of natural phenomena: if it takes a certain amount of force to break a rope, and one applies to the rope a slightly lesser force, what will result is not a partial rupture, that is, the rupture of some part of the strands making up the rope, but merely tension, which is something completely different; if one augments the force in a continuous way, the tension will also increase continuously, but there will come a moment when the rupture will occur, and then, suddenly and as it were instantaneously, there will be an effect of quite another nature than the preceding, which manifestly implies a discontinuity; and thus it is not true to say, in completely general terms and without any sort of restriction, that *natura non facit saltus* [nature does not make leaps].

However that may be, it is at any rate sufficient that geometric magnitudes should be continuous, as indeed they are, in order that one always be able to take from them elements as small as one likes, hence elements that are capable of becoming smaller than any assignable magnitude; and as Leibniz said, 'a rigorous demonstration of the infinitesimal calculus no doubt consists in this,' which applies precisely to these geometric magnitudes. The 'law of continuity' can thus serve as the *fundamentum in re* of these fictions that are the infinitesimal quantities, and, moreover, as well as the other fictions of imaginary roots (since Leibniz linked the two in this respect), but for all that without it being necessary to see in it 'the touchstone of all truth', as he would perhaps have wished. Furthermore, even if one does admit a 'law of continuity', though of course still maintaining certain restrictions as to its range, and even if one recognizes that this law can serve to justify the foundation of the infinitesimal calculus, *modo sano sensu intelligantur*, it by no means follows that one must conceive of it exactly as Leibniz did, or that one must accept all the consequences he attempted to draw from it; it is this conception and these consequences that we must now examine a little more closely.

The Metaphysical Principles of the Infinitesimal Calculus

In its most general form, this law finally amounts to the following, which Leibniz stated on many occasions in different terms, but always with fundamentally the same meaning: whenever there is a certain order to principles understood here in the relative sense of whatever is taken as starting-point, there must always be a corresponding order to the consequences drawn from them. As we have already pointed out, this is then only a particular case of the 'law of justice', or of order, which postulates 'universal intelligibility'. For Leibniz it is therefore fundamentally a consequence or application of the 'principle of sufficient reason', if not this principle itself insofar as it applies more particularly to combinations and variations of quantity. As he says, 'continuity is an ideal thing [which is moreover far from as clear a statement as one might desire], but the real is nevertheless governed by the ideal or abstract . . . because all is governed by reason.'[3] There is assuredly a certain order in things, which is not in question, but this order can be conceived of quite differently from the manner of Leibniz, whose ideas in this regard were always influenced more or less directly by his so-called 'principle of the best', which loses all meaning as soon as one has understood the metaphysical identity of the possible with the real;[4] what is more, although he was a declared adversary of narrow Cartesian rationalism, when it comes to his conception of 'universal intelligibility', one could reproach him for having too readily confused 'intelligible' with 'rational'; but we shall not dwell further on these considerations of a general order, for they would lead us far afield from our subject. In this connection we will only add that one might well be astonished that, after having affirmed that 'mathematical analysis need not depend on metaphysical controversies'—which is quite contestable, moreover, since it amounts to making of mathematics a science entirely ignorant of its own principles, in accordance with the purely profane point of view; besides, incomprehension alone can give birth to controversies in the metaphysical domain—after such an assertion Leibniz himself, in support of his 'law of causality', to which he links this mathematical analysis, finally comes to invoke

3. Previously cited letter to Varignon, February 2, 1702.
4. See *The Multiple States of the Being*, chap. 11.

an argument no longer metaphysical indeed, but definitely theological, which could in turn lead to many other controversies. 'It is because all is governed by reason,' he says, 'and because otherwise there would be neither science nor rules, which would not conform to the nature of the sovereign principle,'[5] to which one could respond that in reality reason is only a purely human faculty, of an individual order, and that, even without having to go back to the 'sovereign principle', intelligence understood in its universal sense, that is, as the pure and transcendent intellect, is something completely different from reason, and cannot be likened to it in any way, such that if it is true that nothing is 'irrational', there are nevertheless many things that are 'supra-rational', but which on that account are no less 'intelligible'.

Let us now move on to a more precise statement of the 'law of continuity', a statement that relates more directly to the principles of the infinitesimal calculus than the preceding: 'If with respect to its data one case approaches another in a continuous fashion and finally disappears into it, it necessarily follows that the results of the cases equally approach in a continuous fashion their sought-out solutions, and that they must finally terminate in one another reciprocally.'[6] There are two things here, which it is important to distinguish: first, if the difference between the two cases diminishes to the point of becoming less than any assignable magnitude *in datis* [in the given], the same must hold *in quaesitis* [in what is sought];

---

5. From the same letter to Varignon.—The first explanation of the 'law of continuity' had appeared in the *Nouvelles de la République des Lettres* in July of 1687, under this rather significant title, and from the same point of view: *Principium quoddam generale non in Mathematicis tantum sed et Physicis utile, cujus ope ex consideratione Sapientiae Divinae examinantur Naturae Leges, qua occasione nata cum R.P. Mallebranchio controversia explicatur, et quidam Cartesianorum errores notantur* [A Certain General Principle, Useful Not Only In Mathematics But In Physics Also, By Which the Laws of Nature are Examined In Reference to Divine Wisdom, and By Which the Controversy Started By R.P. Malebranche Is Explained and Some Errors of the Cartesians Are Pointed Out].

6. *Specimen Dynamicum pro admirandis Naturae Legibus circa corporum vires et mutuas actiones detegendis et ad suas causas revocandis* [A Dynamic Specimen for Studying the Laws of Nature Regarding the Forces of Bodies and Discovering Their Interactions, and For Tracing Their Causes], Part II.

this, in short, is only an application of the more general statement, and this part of the law raises no objections as soon as it is admitted that continuous variations exist and that the infinitesimal calculus is properly linked precisely to the domain in which such variations are effected, namely the geometric domain, but must it be further admitted that *casus in casum tandem evanescat* [one case finally disappears into the other], and that consequently *eventus casuum tandem in se invicem desinant* [the outcomes of the cases finally end in each other]? In other words, will the difference between the two cases ever become rigorously null, in consequence of their continuous and indefinite decrease, or again, if one prefer, will their decrease, though indefinite, ever come to an end? This is fundamentally the question of knowing whether, within a continuous variation, the limit can be reached, and on this point we will first of all make this remark: as the indefinite always includes in a certain sense something of the inexhaustible, insofar as it is implied by the continuous, and as Leibniz moreover did not suppose that the division of the continuous could ever arrive at a final term, nor even that this term could really exist, is it completely logical and coherent on his part to maintain at the same time that a continuous variation, which is effected *per infinitos gradus intermedios* [by infinite intermediary steps],[7] could reach its limit? This is certainly not to say that such a limit can in no way be reached, which would reduce the infinitesimal calculus to no more than a simple method of approximation; but if it is effectively reached, this must not be within the continuous variation itself, nor as a final term in the indefinite sequence of *gradus mutationis* [degrees of change]. Nevertheless, it is by this 'law of continuity' that Leibniz claims to justify the 'passage to the limit', which is not the least of the difficulties to which his method gives rise from the logical point of view, and it is precisely here that his conclusions become completely unacceptable; but to make this aspect of the question entirely understandable, we must begin by clarifying the mathematical notion of the limit itself.

---

7. Letter to Schulenburg, March 29, 1698.

12

The Notion of the Limit

The notion of the limit is one of the most important we have to examine here, for the value of the infinitesimal method, at least insofar as its rigor is concerned, depends entirely upon it; one could even go so far as to say that, ultimately, 'the entire infinitesimal algorithm rests solely on the notion of the limit, for it is precisely this rigorous notion that serves to define and justify all the symbols and formulas of the infinitesimal calculus.'[1] Indeed, the object of this calculus 'amounts to calculating the limits of ratios and the limits of sums, that is, to finding the fixed values toward which the ratios or sums of variable quantities converge, inasmuch as these quantities decrease indefinitely according to a given law.'[2] To be even more precise, let us say that of the two branches into which the infinitesimal calculus may be divided, the differential calculus consists in calculating the limits of ratios, of which the two terms decrease indefinitely, at the same time following a certain law in such a way that the ratio itself always maintains a finite and determined value; and the integral calculus consists in calculating the limits of sums of elements, of which the multitude increases indefinitely as the value of each element decreases indefinitely, for both of these conditions must be united in order for the sum itself always to remain a finite and determined quantity. This being granted, one can say in a general way that the limit of a variable quantity is another quantity considered to be fixed, which the variable quantity is supposed to approach through the values it successively takes on in the course of its variation, until it differs from the fixed quantity

---

1. L. Couturat, *De l'infini mathématique*, introduction, p 23.
2. Ch. de Freycinet, *De l'Analyse infinitésimale*, preface, p 8.

by as little as one likes, or in other words, until the difference between the two quantities becomes less than any assignable quantity. The point which we must emphasize most particularly, for reasons that will be better understood in what follows, is that the limit is essentially conceived as a fixed and determined quantity; even though it will not be given by the conditions of the problem, one should always begin by supposing it to have a determined value, and continue to regard it as fixed until the end of the calculation.

But the conception of the limit in and of itself is one thing, and the logical justification of the 'passage to the limit' quite another; Leibniz believed that

> what in general justifies this 'passage to the limit' is that the same relations that exist among several variable magnitudes also subsist among their fixed limits when their variations are continuous, for then they will indeed reach their respective limits; this is another way of putting the principle of continuity.[3]

But the entire question is precisely that of knowing whether a variable quantity, which approaches its fixed limit indefinitely and which, consequently, can differ from it by as little as one likes, according to the very definition of a limit, can effectively reach this limit precisely as a consequence of this variability, that is, whether a limit can be conceived as the final term in a continuous variation. We shall see that in reality this solution is unacceptable; but putting aside the question, to return to it later, we will only say for now that the true notion of continuity does not allow infinitesimal quantities to be considered as if they could ever equal zero, for they would then cease to be quantities; now, Leibniz himself held that they must always preserve the character of true quantities, even when they are considered to be 'vanishing'. An infinitesimal difference can therefore never be strictly null; consequently, a variable, insofar as it is regarded as such, will always really differ from its limit, and could not reach this limit without thereby losing its variable character.

---

3. L. Couturat, *De l'infini mathématique*, p 268, note.—This is the point of view expressed, notably, in the *Justification du Calcul des infinitésimales par celui de l'Algèbre ordinaire*.

On this point, aside from one slight reservation, we can thus entirely accept the considerations a previously cited mathematician sets forth in these terms:

> What characterizes a limit as we have defined it is that the variable can approach it as much as one might wish, while nonetheless never being able to strictly reach it; for in order that the variable in fact reach it, a certain infinity would have to be realized, which is necessarily ruled out. . . . And one must also keep to the idea of an indefinite, that is to say an even greater, approximation.[4]

Instead of speaking of 'the realization of a certain infinity', which has no meaning for us, we will simply say that a certain indefinitude would have to be exhausted precisely insofar as it is inexhaustible, but that at the same time the possibilities of development contained within this very indefinitude allow the attainment of as close an approximation as might be desired, *ut error fiat minor dato* [that the error may become smaller than any given error], according to an expression of Leibniz, for whom 'the method is certain' as soon as this result is attained.

> The distinctive feature of the limit, and that which prevents the variable from ever exactly reaching it, is that its definition is different from that of the variable; and the variable, for its part, while approaching the limit more and more closely, never reaches it, because it must never cease to satisfy its original definition, which, as we have said, is different. The necessary distinction between the two definitions of the limit and the variable is met with everywhere. . . . This fact, that the two definitions, although logically distinct, are nevertheless such that the objects they define can come closer and closer to one another,[5] explains what

---

4. Ch. De Freycinet, *De l'Analyse infinitésimale*, p18.

5. It would be more exact to say that one of them can come closer and closer to the other since only one of the objects is variable, while the other is essentially fixed; thus, precisely by reason of the definition of the limit, their coming together can in no way be considered to constitute a reciprocal relation, in which the two terms would be as it were interchangeable; moreover, this irreciprocity implies that their difference is of a properly qualitative order.

might at first seem strange, that is, the impossibility of ever making coincide two quantities over which one has the authority to diminish the difference until it becomes so small as to pass beyond expressibility.[6]

There is hardly any need to say that in virtue of the modern tendency to reduce everything exclusively to the quantitative, some people have not failed to find fault with this conception of the limit for introducing a qualitative difference into the science of quantity itself; but if it must be discarded for this reason, it would likewise be necessary to ban from geometry entirely—among other things—the consideration of similarity, which is also purely qualitative, since it concerns only the form of figures, abstracting them from their magnitudes, and hence from their properly quantitative element, as we have already explained elsewhere. In this connection, it would also be good to note that one of the chief uses of the differential calculus is to determine the directions of the tangents at each point on a curve, the totality of which defines the very form of the curve, and that in the spatial order direction and form are precisely elements of an essentially qualitative character.[7] What is more, it is no solution to claim to purely and simply do away with the 'passage to the limit' on the pretext that the mathematician can dispense with actually passing to it without in any way hindering him from pushing his calculation to its end; this may be true, but what matters is this: under these conditions, up to what point would one have the right to consider this calculus to rest on rigorous reasoning, and even if 'the method is thus certain', will it not be so only as a simple method of approximation? One could object that the conception we just explained also makes the 'passage to the limit' impossible, since the character of this limit is precisely such as to prevent its ever being reached; but this is true only in a certain sense, and only insofar as one considers variable quantities as such, for we did not say that the limit could in no way be reached, but—and it is essential that this be made clear—that it could not be reached within the

6. Ibid., p19.
7. See *The Reign of Quantity and the Signs of the Times*, chap. 4.

variation, and as a term of the latter. The only true impossibility is the notion of a 'passage to the limit' constituting the result of a continuous variation; we must therefore replace it with another notion, and this we shall do more explicitly in what follows.

13

Continuity and
Passage to the Limit

We can now return to our examination of the 'law of continuity', or, to be more exact, to the aspect of the law that we had to momentarily lay aside, and which is precisely that aspect by which Leibniz believed 'passage to the limit' could be justified. For him what follows from it is

> that with continuous quantities, the extreme exclusive case may be treated as inclusive, and that such a case, although totally different in nature, is thus as if contained in a latent state in the general law of the other cases.[1]

Although Leibniz himself does not appear to have suspected it, it is precisely here that the principal logical error in his conception of continuity lies, which one may quite easily recognize in the consequences he draws from it and in the ways in which he applies it. Here are a few examples:

> In accordance with my law of continuity, one is allowed to consider rest to be an infinitely small motion, that is, to be equivalent to a species of its contradictory, and coincidence to be an infinitely small distance, equality the last of inequalities, etc.[2] [Or again]: In accordance with this law of continuity, which excludes all sudden changes, the case of rest can be regarded as a

---

1. *Epistola ad V. Cl. Christianum Wolfium, Professorem Matheseos Halensem, circa Scientiam Infiniti* [Letter to V. Cl. Christian Wolf, Mathematics Professor Halensem, concerning the Science of the Infinite], in the *Acta Eruditorum* of Leipzig, 1713.
2. Previously cited letter to Varignon, February 2, 1702.

special case of motion, namely as a vanishing or minimum motion, and the case of equality as a case of vanishing inequality. It follows that the laws of motion must be established in such a way that there be no need for special rules for bodies in equilibrium and at rest, but that the latter should themselves arise from the rules concerning bodies in disequilibrium and in motion; or, if one does wish to set forth particular rules for rest and equilibrium, one must take care that they not be such as to disagree with the hypothesis that holds rest to be an incipient motion or equality the final inequality.[3]

Let us add one more quotation on the subject, in which we find a new example, of a somewhat different kind from the preceding, but no less contestable from the logical point of view:

Although it is not rigorously true that rest is a species of motion, or that equality is a species of inequality, just as it is not true that the circle is a species of regular polygon, one can nevertheless say that rest, equality, and the circle are the terminations of motion, inequality, and the regular polygon, which, by continual change, arrive at the former by vanishing. And although these terminations are exclusive, that is, not rigorously included within the varieties they limit, they nevertheless have the same properties as they would if they were so included, in accordance with the language of infinites or infinitesimals, which takes the circle, for example, as a regular polygon with an infinite number of sides. Otherwise the law of continuity would be violated, that is to say that because one passes from polygons to the circle by a continual change, without any break, there must likewise be no break in the passage from the attributes of polygons to those of the circle.[4]

It is worth pointing out that, as is indicated in the beginning of the last passage cited above, Leibniz considers these assertions to be

3. *Specimen Dynamicum*, previously cited above.
4. *Justification du Calcul des infinitésimales par celui de l'Algèbre ordinaire*, note added to the letter from Varignon to Leibniz of May 23, 1702, in which it is mentioned as having been sent by Leibniz to be inserted in the *Journal de Trévoux*. Leibniz takes the word 'continual' in the sense of 'continuous'.

of the same kind as those that are merely *toleranter verae*, which, he says elsewhere,

> above all serve the art of invention, although, in my opinion, they contain something of the fictional and imaginary which however can easily be rectified by reducing them to ordinary expressions, in order that they not produce error.[5]

But are they not precisely that already, and in reality do they not rather contain contradictions pure and simple? No doubt Leibniz recognized that the extreme case, or *ultimus casus*, is *exclusivus*, which obviously implies that it falls outside of the series of cases that are naturally included in the general law; but then with what right can it be included in this law in spite of it, and be treated *ut inclusivum* [as inclusive], that is, as if it were only one particular case contained within the series? It is true that the circle is the limit of a regular polygon with an indefinitely increasing number of sides, but its definition is essentially other than that of polygons; and in such an example one can see quite clearly that there exists a qualitative difference between the limit itself and that of which it is the limit, as we have said before. Rest is in no way a particular case of motion, nor equality a particular case of inequality, nor coincidence a particular case of distance, nor parallelism a particular case of convergence; besides, Leibniz does not suppose that they are so in a rigorous sense, but he nonetheless maintains that they can in some way be regarded as such, with the result that 'the genus terminates in the opposed quasi-species,' and that something can be 'equivalent to a species of its contradictory.'[6] Moreover, let us note in passing that Leibniz's notion of 'virtuality' seems to be linked to this same order of ideas, as he gives it the special sense of potentiality viewed as incipient actuality,[7] which again is no less contradictory than the other examples just cited.

5. 'Epistola ad V. Cl. Christianum Wolfium', previously cited above.

6. *Initia Rerum Mathematicarum Metaphysica* [The Metaphysical Principles of Mathematicals]. Leibniz's exact words are *genus in quasi-speciem oppositam desinit*, and use of the singular expression 'quasi-species' seems at the very least to indicate a certain difficulty in giving a more plausible appearance to such a statement.

7. The words 'actuality' and 'potentiality' are of course taken here in their Aristotelian and Scholastic sense.

Whatever the point of view from which things are envisaged, it is not in the least clear that a certain species could be a 'borderline case' of the opposite species or genus, for it is not in this way that opposed things limit each other reciprocally, but definitely to the contrary in that they exclude one another, and it is impossible for one contradictory to be reduced to another; for example, can inequality have any significance apart from the degree to which it is opposed to equality and is its negation? We certainly cannot say that assertions such as these are even *toleranter verae*, for even if one does not accept the existence of absolutely separate genuses, it is nonetheless true that any genus, defined as such, can never become an integral part of another equally defined genus when the definition of this latter does not include its own, even if it does not exclude it formally as in the case of contradictories; and if a connection can be established between different genuses, this is not in virtue of that in which they effectively differ, but only in virtue of a higher genus, which includes both. Such a conception of continuity, which ends up abolishing not only all separation, but even all effective distinction, in allowing direct passage from one genus to another without reducing the two to a higher or more general genus, is in fact the very negation of every true logical principle; and from this to the Hegelian affirmation of the 'identity of contradictories' is then but one step which is all too easy to take.

14

Vanishing Quantities

For Leibniz, the justification for 'passage to the limit' ultimately consists in the fact that the particular case of the 'vanishing quantities', as he says, must in a certain sense be included within the general rule by virtue of continuity; moreover, these vanishing quantities cannot be regarded as 'absolute nothings', or as pure zeros, for by reason of the same continuity they maintain among themselves determined ratios—and generally differ from unity—in the very instant in which they vanish, which implies that they are still real quantities, although 'unassignable' with respect to ordinary quantities.[1] However, if these vanishing quantities—or the infinitesimal quantities, which amounts to the same thing—are not 'absolute nothings', even when it is a question of differentials of orders higher than the first, they must still be considered 'relative nothings', which is to say that, while retaining the character of real quantities, they can and must be negligible with regard to ordinary quantities, with which they are 'incomparable';[2] but multiplied by 'infinite' quantities, or quantities incomparably greater than ordinary ones, they again produce these ordinary quantities, which could not be so if they were absolutely nothing. In light of the definitions we presented earlier, one can see that the consideration of the ratios of

1. For Leibniz, $0/0 = 1$, since, as he says, 'one nothing is the same as another'; but as $(0)(n)$ is also equal to 0, for any value of n, it is obvious that one could just as well write $0/0 = n$, and this is why the expression $0/0$ is generally thought of as representing what is called an 'indeterminate form'.

2. The difference between this and the comparison of the grain of sand is that as soon as one speaks of 'vanishing quantities', it is necessarily a question of variable quantities, and no longer of fixed and determined quantities, however small one might suppose them to be.

vanishing but still determined quantities refers to the differential calculus, while the consideration of the multiplication of these quantities by 'infinite' quantities, yielding ordinary quantities, refers to the integral calculus. The difficulty in all this is to admit that quantities that are not absolutely null must nonetheless be treated in the calculus as if they were, which risks giving the impression that it is merely a question of simple approximation; again, in this regard Leibniz sometimes seems to invoke the 'law of continuity', by which the 'borderline case' finds itself included within the general rule, as if this were the only postulate his method required; this argument is quite unclear, however, and one should rather return to the notion of 'incomparables', as he himself often does, moreover, in order to justify the elimination of infinitesimal quantities from the results of the calculus.

Indeed, Leibniz considers as equal not only those quantities of which the difference is null, but even those of which the difference is incomparable with respect to the quantities themselves; this notion of 'incomparables' is, for him, the foundation not only for the elimination of infinitesimal quantities, which thus disappear in the face of ordinary quantities, but also for the distinction between different orders of infinitesimal or differential quantities, the quantities of each order being incomparable with respect to those of the preceding, as those of the first order are with respect to ordinary quantities, but without ever arriving at 'absolute nothings'. 'I call two magnitudes incomparable,' says Leibniz, 'when one, despite multiplication by any finite number whatsoever, can nonetheless not exceed the other, in the same way that Euclid treated it in the fifth definition of his fifth book.'[3] However, there is nothing there to indicate whether this definition should be understood of fixed and determined, or of variable, quantities; but one can admit that in all its generality it must apply without distinction to both cases; the entire question would then be one of knowing whether two fixed quantities, however different they might be within the scale of magnitudes, could ever be regarded as truly 'incomparable', or whether they would only be so relative to the means of measurement at our disposal. But we

---

3. Letter to Marquis de l'Hospital, June 14–24, 1695.

shall not dwell further on this point, since Leibniz himself declared elsewhere that this is not the case with differentials,[4] from which it is necessary to conclude, not only that the comparison of the grain of sand is in itself manifestly faulty, but also that it fundamentally does not answer, even in his own thought, to the true notion of 'incomparables', at least insofar as this notion must be applied to the infinitesimal quantities.

Some people, however, have believed that the infinitesimal calculus can be rendered perfectly rigorous only on the condition that the infinitesimal quantities be regarded as null, and at the same time they have wrongly thought that one can suppose an error to be null as long as one can also suppose it to be as small as one likes; wrongly, we say, for that would be the same as to admit that a variable, as such, could reach its limit. Here is what Carnot has to say on the subject:

> There are those who believe they have sufficiently established the principle of infinitesimal analysis with the following reasoning: it is obvious, they say, and universally acknowledged, that the errors to which the procedure of infinitesimal analysis would give rise—if there were any—could always be supposed as small as one might wish; it is also obvious that any error one is free to suppose as small as one likes is null, for since one can suppose it to be as small as one wishes, one can suppose it to be zero; therefore, the results of the infinitesimal analysis are rigorously exact. This argument, plausible at first sight, is nevertheless anything but valid, for it is false to say that because one is free to render an error as small as one likes one can thus render it absolutely null. . . . One is faced with the necessary alternative either of committing an error, however slight one might suppose it to be, or of falling back on a formula that says nothing, and such is precisely the crux of the difficulty with the infinitesimal analysis.[5]

It is certain that any formula in which a ratio appears in the form $^0/_0$ 'says nothing', and one could even say that it has no meaning in

4. Previously cited letter to Varignon, February 2, 1702.
5. *Réflexions sur la Métaphysique du Calcul infinitésimal,* p36.

80

and of itself; it is only in virtue of a convention—justified, moreover —that one can give any sense to the expression $^0/_0$, regarding it as a symbol of indeterminacy;[6] but this very indeterminacy then means that the ratio in this form can be equal to anything, whereas on the contrary it must maintain a determined value in every particular case; it is the existence of this determined value that Leibniz puts forward,[7] and in itself this argument is completely unassailable.[8] However, it is quite necessary to recognize that the notion of 'vanishing quantities' has 'the tremendous drawback of considering quantities in that state in which they so to speak cease to be quantities', to use Lagrange's expression; but contrary to what Leibniz thought, there is no need to consider them precisely in the instant in which they vanish, nor even to suppose that they really could vanish, for in that case they would indeed cease to be quantities. Moreover, this essentially supposes that strictly speaking there is no 'infinitely small' quantity, for this 'infinitely small' quantity—or at least what would be called such in Leibniz's language—could only be zero, just as an 'infinitely great' quantity, taken in the same sense, could only be an 'infinite number'; but in reality zero is not a number, and 'null quantities' have no more existence than do 'infinite quantities'. The mathematical zero, in its rigorous and strict sense, is but a negation, at least as far as its quantitative aspect is concerned, and one cannot say that the absence of quantity itself constitutes a quantity; we shall return to this point shortly, in order to develop more completely the consequences that result from it.

In sum, the expression 'vanishing quantities' has above all the drawback of producing an equivocation, and of leading to the belief that infinitesimal quantities can be considered as quantities that are

---

6. On this subject, see the preceding note.

7. With this difference, namely that for him the ratio $^0/_0$ is not indeterminate, but always equal to 1, as we pointed out earlier, whereas in fact the value in question differs in each case.

8. Cf. Ch. de Freycinet, *De l'Analyse infinitésimale*, pp 45–46: 'If the increases are reduced to the state of pure zeros, they will no longer have any meaning. Their property is to be, not rigorously null, but indefinitely decreasing, without ever being confounded with zero, in virtue of the general principle that a variable can never coincide with its limit.'

effectively annulled, for without altering the meaning of these words, it is difficult to understand how, when it is a question of quantities, 'to vanish' could mean anything other than to be annulled. In reality, these infinitesimal quantities, understood as indefinitely decreasing quantities, which is their true significance, can never be called 'vanishing' in the proper sense of the word. It would most certainly have been preferable had the notion never been introduced, as it is fundamentally bound up with Leibniz's conception of continuity, and, as such, inevitably contains the same element of contradiction inherent in the illogicality of this latter. Now if an error, despite being able to be rendered as small as one likes, can never become absolutely null, how can the infinitesimal calculus be truly rigorous, and if the error is in fact only practically negligible, would it not be necessary to conclude that the calculus is thus reduced to a simple method of approximation, or at least, as Carnot says, of 'compensation'? This is a question that we must resolve in what follows; but as we have here been brought to speak of zero and of the so-called 'null quantity', it will be worthwhile to deal with this other subject first, the importance of which, as we shall see, is far from negligible.

15

Zero is Not a Number

The indefinite decrease of numbers can no more end in a 'null number' than their indefinite increase can in an 'infinite number', and for the same reason, since each of these numbers must be the inverse of the other; indeed, in accordance with what was said earlier on the subject of inverse numbers, as each of the two sets—the one increasing, the other decreasing—is equally distant from the unit, the common point of departure for both, and as there must further necessarily be as many terms in the one as in the other, their final terms—namely, the 'infinite number' and the 'null number'—if they existed, would themselves have to be equally distant from the unit, and thus the inverses of one another.[1] Under these conditions, if the sign ∞ is in reality only a symbol for indefinitely increasing quantities, then logically the sign o should likewise be able to be taken as a symbol for indefinitely decreasing quantities, in order to express in notation the symmetry that, as we have said, exists between the two; but unfortunately this sign o already has quite another significance, for it originally served to designate the complete absence of quantity, whereas the sign ∞ has no real sense that would correspond to the former. Here, as with the 'vanishing quantities', we

---

1. In ordinary notation this would be represented by the formula $(o)(\infty) = 1$; but in fact the form $^o\!/\!\infty$ is again, like $^o\!/\!o$, an 'indeterminate form', and one could write $(o)(\infty) = n$, where n stands for any number, which moreover shows that, in reality, o and ∞ cannot be regarded as representing determined numbers; we shall return to this point later. In another respect, one could remark that $(o)(\infty)$ is for the 'limits of sums' of the integral calculus what $^o\!/\!o$ is for the 'limits of ratios' of the differential calculus.

have yet another source of confusion, and in order to avoid this it would be necessary to create another symbol, apart from zero, for indefinitely decreasing quantities, since these quantities are characterized precisely by the fact that they can never be annulled, despite any variation they might undergo; at any rate, with the notation currently employed by mathematicians, it seems almost impossible to prevent confusions from arising.

If we emphasize the fact that zero, insofar as it represents the complete absence of quantity, is not a number and cannot be considered as such—even though this might appear obvious enough to those who have never had occasion to take cognizance of certain disputes—this is because, as soon as one admits the existence of a 'null number', which would have to be the 'smallest of numbers', one is inevitably led by way of correlation to suppose as its inverse an 'infinite number', in the sense of the 'greatest of numbers'. If, therefore, one accepts the postulate that zero is a number, the arguments in favor of an 'infinite number' follow in a perfectly logical manner;[2] but it is precisely this postulate that we must reject, for if the consequences deduced from it are contradictory—and we have seen that the existence of an 'infinite number' is indeed so—then the postulate in itself must already imply contradiction. Indeed, the negation of quantity can in no way be assimilated to a particular quantity; the negation of number or of magnitude can in no sense and to no degree constitute a species of number or magnitude; to claim the contrary would be to maintain that a thing could be 'equivalent to a species of its contradictory,' to use Leibniz's expression, and would be as much as to say immediately that the negation of logic is itself logic.

It is therefore contradictory to speak of zero as a number, or to suppose that a 'zero in magnitude' is still a magnitude, from which would inevitably result the consideration of as many distinct zeros as there are different kinds of magnitude; in reality, there can only be zero pure and simple, which is none other than the negation of

---

2. Indeed, the arguments of L. Couturat in his thesis *De l'infini mathématique* rest, in large part, on this postulate.

quantity, whatever the mode envisaged.[3] When such is accepted as the true sense of the arithmetical zero, taken 'rigorously', it becomes obvious that this sense has nothing in common with the notion of indefinitely decreasing quantities, which are always quantities; they are never an absence of quantity, nor again are they anything that is as it were intermediate between zero and quantity, which would be yet another completely unintelligible conception, and which in its own order would recall that of Leibnizian 'virtuality', which we had occasion to mention earlier.

We can now return to the other meaning that zero actually has in common notation, in order to see how the confusions we spoke of were introduced. We said earlier that in a way a number can be regarded as practically indefinite when it is no longer possible for us to express or represent it distinctly in any way; such a number, whatever it might be, can only be symbolized in the increasing order by the sign ∞, insofar as this represents the indefinitely great; it is therefore not a question of a determined number, but rather of an entire domain, and this is necessary moreover if it is to be possible to envisage inequalities and even different orders of magnitude within the indefinite. Mathematical notation lacks a symbol for the corresponding domain in the decreasing order, what might be called the domain of the indefinitely small; but since a number belonging to this domain is, in fact, negligible in calculations, it is in practice habitually considered to be null, even though this is only a simple approximation resulting from the inevitable imperfection of our means of expression and measurement, and it is doubtless for this reason that it came to be represented by the same symbol o that also represents the rigorous absence of quantity. It is only in this

---

3. From this it further results that zero cannot be considered a limit in the mathematical sense of the word, for by definition a true limit is always a quantity; moreover, it is evident that a quantity that decreases indefinitely has no more of a limit than does a quantity that increases indefinitely, or at least that neither can have any other limits than those that necessarily result from the very nature of quantity as such, which is a rather different use of the word 'limit', although there is a certain connection between the two meanings, as will be shown later; mathematically, one can speak only of the limit of the ratio of two indefinitely increasing or indefinitely decreasing quantities, and not of the limit of these quantities themselves.

sense that the sign 0 becomes in a way symmetrical to the sign ∞ and that the two can be placed respectively at the two extremities of the sequence of numbers as we envisaged it earlier, with the whole numbers and their inverses extending indefinitely in the two opposite directions of increase and decrease. This sequence then presents itself in the following form: $0 \ldots \frac{1}{4}, \frac{1}{3}, \frac{1}{2}, 1, 2, 3, 4 \ldots \infty$; but we must take care to recall that 0 and ∞ represent not two determined numbers terminating the series in either direction, but two indefinite domains, in which on the contrary there can be no final terms, precisely by reason of their indefinitude; moreover, it is obvious that here zero can be neither a 'null number', which would be a final term in the decreasing direction, nor again a negation or absence of quantity, which would have no place in this sequence of numerical quantities.

As we explained previously, any two numbers in the sequence that are equidistant from the central unit are inverses or complementaries of one another, thus producing the unit when multiplied together: $(1/n)(n) = 1$, such that for the two extremities of the sequence, one would be led to write $(0)(\infty) = 1$ as well; but, since the signs 0 and ∞, the two factors of this product, do not represent determined numbers, it follows that the expression $(0)(\infty)$ itself constitutes a symbol of indeterminacy, or what one would call an 'indeterminate form', and one must therefore write $(0)(\infty) = n$, where n could be any number;[4] it is no less true that in any case one will thus be brought to ordinary finitude, the two opposed indefinites so to speak neutralizing one another. Here, once again, one can clearly see that the symbol ∞ most emphatically does not represent the Infinite, for the Infinite, in its true sense, can have neither opposite nor complementarity, nor can it enter into correlation with anything at all, no more with zero, in whatever sense it might be understood, than with the unit, or with any number, or again with any particular thing of any order whatsoever, quantitative or not; being the absolute and universal All, it contains Non-Being as well as Being, such that zero itself, whenever it is not regarded as

4. On this subject, see the preceding note.

purely nothing, must also necessarily be considered to be contained within the Infinite.

In alluding here to Non-Being, we touch on another meaning of zero quite different from those we have just considered, the most important from the point of view of metaphysical symbolism; but in this regard, in order to avoid all confusion between the symbol and that which it represents, it is necessary to make it quite clear that the metaphysical Zero, which is Non-Being, is no more the zero of quantity than the metaphysical Unit, which is Being, is the arithmetical unit. What is thus designated by these terms is so only by analogical transposition, since as soon as one places oneself within the Universal one is obviously beyond every special domain such as that of quantity. Furthermore, it is not insofar as it represents the indefinitely small that zero by such a transposition can be taken as a symbol of Non-Being, but, following its most rigorous mathematical usage, rather insofar as it represents the absence of quantity, which in its order indeed symbolizes the possibility of non-manifestation, just as the unit, since it is the point of departure for the indefinite multiplicity of numbers, symbolizes the possibility of manifestation as Being is the principle of all manifestation.[5]

This again leads us to note that zero, however it may be envisaged, can in no case be taken for pure nothingness, which corresponds metaphysically only to impossibility, and which in any case cannot logically be represented by anything. This is all too obvious when it is a question of the indefinitely small; it is true that this is only a derivative sense, so to speak, due, as we were just saying, to a sort of approximate assimilation of quantities negligible for us, to the total absence of quantity; but insofar as it is a question of this very absence of quantity, what is null in this connection certainly cannot be so in other respects, as is apparent in an example such as the point, which, being indivisible, is by that very fact without extension, that is, spatially null,[6] but which, as we have explained

---

5. On this subject, see *The Multiple States of the Being*, chap. 3.

6. This is why the point can in no way be considered as constituting an element or part of length, as we said earlier.

elsewhere, is nonetheless the very principle of all extension.[7] It is quite strange, moreover, that mathematicians are generally inclined to envisage zero as a pure nothingness, when it is nevertheless impossible for them not to regard it at the same time as endowed with an indefinite potentiality, since, placed to the right of another digit termed 'significant', it contributes to forming the representation of a number that, by the repetition of this same zero, can increase indefinitely, as is the case with the number ten and its successive powers for example. If zero were really only pure nothingness, this could not be so; and indeed, in that case it would only be a useless sign, entirely deprived of effective value; here we have yet another inconsistency to add to the list of those that we have already had occasion to point out in the conceptions of modern mathematicians.

7. See *The Symbolism of the Cross*, chap. 16.

16

The Notation of
Negative Numbers

If we now return to the second and more important of the two mathematical senses of zero, namely that of zero considered as a representation of the indefinitely small, this is because within the doubly indefinite sequence of numbers the domain of the latter embraces all that eludes our means of evaluation in a certain direction, just as within the same sequence the domain of the indefinitely great embraces all that eludes these means of evaluation in the other direction. This being so, to speak of numbers 'less than zero' is obviously no more appropriate than to speak of numbers 'greater than the indefinite', and it is all the more unacceptable—if such is possible—when zero is taken in its other sense as purely and simply representing the absence of quantity, for it is totally inconceivable that a quantity should be less than nothing. In a certain sense, however, this is precisely what is done when one introduces the consideration of so-called negative numbers to mathematics, forgetting as a result of modern 'conventionalism' that these numbers were originally no more than an indication of the result of a subtraction that is in fact impossible, in which a greater number is taken away from a smaller; besides, we have already pointed out that all generalizations or extensions of the idea of number arise only from the consideration of operations that are impossible from the point of view of pure arithmetic; but this conception of negative numbers, and the consequences it entails, demand some further explanation.

We said earlier that the sequence of whole numbers is formed starting from the unit, and not from zero; indeed, the unit being fixed, the entire sequence of numbers is inferred from it in such a way that one could say that it is already implied and contained in principle within the initial unit[1] whereas it is obvious that no number can be derived from zero. Passage from zero to the unit cannot be made in the same way as passage from the unit to other numbers, or from any given number to the next, and to suppose the passage from zero to the unit possible is to have already implicitly posited the unit.[2] Finally, to place zero at the beginning of the sequence of numbers as if it were the first in the sequence, can mean only one of two things: either one admits, contrary to what has already been established, that zero really is a number, and consequently that its ratios with respect to other numbers are of the same order as the ratios of these numbers are to each other—which is not the case, since zero multiplied or divided by a given number is always zero— or this is a simple device of notation, which can only lead to more or less inextricable confusions. In fact, the use of this device is never justified except to permit the introduction of the notation of negative numbers, and if such notation doubtless offers certain advantages for the convenience of calculation—an entirely 'pragmatic' consideration, which is not in question here and which is even without any real importance from our point of view—it is easy to see that it is not without grave logical difficulties. The first of these is precisely the conception of negative quantities as 'less than zero', an affirmation which Leibniz ranked among the affirmations that are only *toleranter verae*, but which in reality is, as we were just saying, entirely devoid of meaning. 'To affirm an isolated negative quantity as less than zero,' says Carnot, 'is to veil the science of mathematics, which should be a science of the obvious, in an impenetrable cloud, and to thrust oneself into a labyrinth of paradoxes, each more bizarre than the last.'[3] On this point we may follow his judgment,

---

1. Similarly, by analogous transposition, all the indefinite multiplicity of the possibilities of manifestation is contained, 'eminently' and in principle, within pure Being, or the metaphysical Unit.

2. This will appear completely obvious if, in conformity with the general law of formation for the sequence of numbers, one represents this passage by the formula $0 + 1 = 1$.

3. 'Note sur les quantités négatives', placed at the end of the *Réflexions sur la Métaphysique du Calcul infinitésimal*, p173.

which is above suspicion and is certainly not exaggerated; moreover, one should never forget in using this notation of negative numbers that it is a matter of nothing more than a simple convention.

The reason for this convention is as follows: when a given subtraction is arithmetically impossible, its result is nonetheless not devoid of meaning when this subtraction is linked to magnitudes that can be reckoned in two opposite directions, as, for example, with distances measured on a line, or angles of rotation around a fixed point, or again the time elapsed in moving from a certain instant toward either the past or the future. From this results the geometric representation habitually accorded negative numbers: taking an entire straight line, indefinite in both directions, and not in one only, as was the case earlier, the distances along the line are considered positive or negative depending on whether they fall one way or the other, and a point is chosen to serve as the origin, in relation to which the distances are positive on one side and negative on the other. For each point on the line there is a number corresponding to the measurement of its distance from the origin, which, in order to simplify our language, we can call its coefficient; once again, the origin itself will naturally have zero for its coefficient, and the coefficients of all the other points on the line will be numbers modified by the signs + and –, which in reality simply indicate on which side the point falls in relation to the origin. On a circumference one could likewise designate positive and negative directions of rotation, and starting from an initial position of the radius, one would take each angle to be positive or negative according to the direction in which it lies, and so on analogously. But to keep to the example of the straight line, two points equidistant from the origin, one on either side, will have the same number for their coefficients, but with contrary signs, and in all cases, a point that is further than another from the origin will naturally have a greater coefficient; thus it is clear that if a number n is greater than another number m, it would be absurd to say, as is ordinarily done, that $-n$ is smaller than $-m$, since on the contrary it represents a greater distance. Moreover the sign thus placed in front of a number cannot really modify it in any way with regard to quantity, since it represents nothing with respect to the measurements of distances themselves, but only the direction in which these distances are traversed, which, properly speaking, is an element of a qualitative, and not a quantitative, order.[4]

Moreover, as the line is indefinite in both directions, one is led to envisage both a positive and a negative indefinite, represented by the signs ∞ and $-\infty$ respectively, commonly designated by the absurd expressions 'greater infinity' and 'lesser infinity'. One might well ask what a negative infinity would be, or again what could remain were one to take away an infinite amount from something, or even from nothing, since mathematicians regard zero as nothing; one has only to put these matters in clear language in order to see immediately how devoid of meaning they are. We must further add that particularly when studying the variation of functions, one is then led to believe that the negative and the positive indefinite merge in such a way that a moving object departing from the origin and moving further and further away in the positive direction would return to the origin from the negative side, or inversely, if the movement were followed for an indefinite amount of time, whence it would result that the straight line, or what would then be considered as such, would in reality be a closed line, albeit an indefinite one. Furthermore, one could show that the properties of a straight line in a plane would be entirely analogous to those of a great circle, or diametrical circle on the surface of a sphere, and that the plane and the straight line could thus be likened respectively to a sphere and a circle of indefinitely great radius, and consequently of indefinitely small curvature, ordinary circles in the plane then being comparable to the smaller circles on the sphere; for this analogy to be rigorous, one would further have to suppose a 'passage to the limit', for it is obvious that however great a radius might become through indefinite increase, it always describes a sphere and not a plane, and that the sphere only tends to be merged with the plane, and its great

4. See *The Reign of Quantity and the Signs of the Times*, chap. 4. One might wonder whether there is not to be found some sort of unconscious memory, as it were, of this qualitative character in the fact that mathematicians still sometimes designate numbers taken 'with their sign', that is, considered to be positive or negative, by the name of 'qualified numbers', although they otherwise do not seem to attach any very clear meaning to this expression.

circle [or diameter] with lines, such that plane and line are limits, in the same way that a circle is the limit of a regular polygon with an indefinitely increasing number of sides. Without pushing the issue further, we shall only remark that through considerations of this sort, one can as it were directly grasp the precise limits of spatial indefinitude; how then, if one wishes to maintain some appearance of logic, can one still speak of the infinite in all this?

When considering positive and negative numbers as we have just done, the sequence of numbers takes the following form: $-\infty \ldots -4$, $-3, -2, -1, 0, 1, 2, 3, 4 \ldots +\infty$, the order of these numbers being the same as that of the corresponding points on the line, that is, the points having these numbers for their respective coefficients, which, moreover, is the mark of the real origin of the sequence thus formed. Although the sequence is equally indefinite in both directions, it is completely different from the one we envisaged earlier, which contained the whole numbers and their inverses: this one is symmetric not with respect to the unit, but with respect to zero, which corresponds to the origin of the distances; and if two numbers equidistant from this central term are to return to it, it will not be by multiplication, as in the case of inverse numbers, but by 'algebraic' addition, that is, effected while taking account of signs, which in this case would amount to a subtraction, arithmetically speaking. Moreover, we can by no means say of the new sequence that it is indefinitely increasing in one direction and indefinitely decreasing in the other, as we could of the preceding, or at least, if one claims to consider it thus, this is only a most incorrect 'manner of speaking', as is the case when one envisages numbers 'less than zero'. In reality, the sequence increases indefinitely in both directions equally, since it is the same sequence of whole numbers that is contained on either side of the central zero; what is called the 'absolute value'—another rather singular expression—must only be taken into consideration in a purely quantitative respect, the positive or negative signs changing nothing in this regard, since, in reality, they express no more than differences in 'situation', as we have just explained. The negative indefinite is therefore by no means comparable to the indefinitely small; on the contrary, it belongs with the indefinitely great as does the positive indefinite; the only difference, which is not one of a quantitative order, is that it proceeds in another direction, which is perfectly conceivable when it is a question of spatial or temporal magnitudes, but totally devoid of meaning for arithmetical magnitudes, for which such a progression is necessarily unique since it cannot be anything other than that of the very sequence of whole numbers.

Among the bizarre or illogical consequences of the notation of negative numbers, we shall further draw attention to the consideration of so-called 'imaginary' quantities which were introduced in the solving of algebraic equations and which, as we have seen, Leibniz ranked at the same level as infinitesimal quantities, namely as what he called 'well-founded fictions'. These quantities, or what are so called, are presented as the roots of negative numbers, although in reality this again only corresponds to a pure and simple impossibility, since, whether a number is positive or negative, its square is necessarily always positive by virtue of the rules of algebraic multiplication. Even if one could manage to give these 'imaginary' quantities some other meaning, thereby making them correspond to something real—a possibility we shall not examine here—it is nonetheless quite certain that their theory and application to analytic geometry as it is presented by contemporary mathematicians never appears as anything but a veritable web of confusions and even absurdities, and as the product of a need for excessive and entirely artificial generalizations, which need does not retreat even before manifestly contradictory propositions; certain theorems concerning the 'asymptotes of a circle', for example, amply suffice to prove that this remark is by no means exaggerated. It is true that one could say that this is no longer a question of geometry properly speaking, but, like the consideration of a 'fourth dimension' of space,[5] only of algebra translated into geometric language; but precisely because such a translation, as well as its inverse, is possible and legitimate to a certain degree, some people would also like to extend it to cases where it can no longer mean anything, and this is indeed quite serious, for it is the symptom of an extraordinary confusion of ideas, as well as the extreme result of a 'conventionalism' taken so far as to cause some people to lose all sense of reality.

5. Cf. *The Reign of Quantity and the Signs of the Times*, chaps. 18 and 23.

17

Representation of the Equilibrium of Forces

In connection with negative numbers, we shall now speak of the rather disputable consequences of the use of these numbers from the point of view of mechanics, even though this is only a digression with respect to the principal subject of our study; moreover, since in virtue of its object the field of mechanics itself is in reality a physical science, the very fact that it is treated as an integral part of mathematics in consequence of the exclusively quantitative point of view of science today means that some rather singular distortions have been introduced. Let us only say that the so-called 'principles' upon which modern mathematicians would build this science, such as they conceive of it, can be referred to as 'principles' only in a completely abusive manner, as they are in fact only more or less well-founded hypotheses, or again, in the most favorable case, only simple laws that are general to some degree, perhaps more general than others, if one likes, but still having nothing in common with true universal principles; in a science constituted according to the traditional point of view, the laws of mechanics would at most be mere applications of these principles to an even more specialized domain. Without entering into excessively lengthy explanations, let us cite as an example of the first case, the so-called 'principle of inertia', which nothing can justify, neither experience, which on the contrary shows that inertia has no role in nature, nor in the understanding, which cannot conceive of this so-called inertia that consists only in a complete absence of properties; one could only legitimately apply such a word to the pure potentiality of universal substance, or to the *materia prima* of the Scholastics, which is moreover for this very reason properly 'unintelligible'; but this *materia prima* is assuredly something completely different from the 'matter' of the physicists.[1] An example of the second case may be seen in what is called the 'principle of the equality of action and reaction', which is so little a principle that it is immediately deduced from the general law of the equilibrium of natural forces: whenever this equilibrium is disturbed in any way, it immediately tends to re-establish itself, whence a reaction of which the intensity is equivalent to that of the action that provoked it. It is therefore only a simple, particular case of what the Far-Eastern tradition calls 'concordant actions and reactions', a principle that does not concern the corporeal world alone, as do the laws of mechanics, but indeed the totality of manifestation in all its modes and states; and for a moment we propose to dwell precisely on this question of equilibrium and its mathematical representation, for it is important enough in itself to merit a momentary pause.

Two forces in equilibrium are usually represented by two opposed 'vectors', that is, by two line segments of equal length, but aimed in opposite directions: if two forces applied to the same point have the same intensity and fall along the same line, but in opposite directions, they are in equilibrium; as they are then without action at their point of application, it is even commonly said that they cancel each other out, although this ignores the fact that if one of the forces is suppressed, the other will immediately act, which proves that they were never really cancelled in the first place. The forces are characterized by numerical coefficients proportional to their respective intensities, and two forces of opposing direction are given coefficients with different signs, the one positive, the other negative, so that if the one is f, the other will be $-f'$. In the case we have just considered, in which the two forces are of the same intensity, the coefficients characterizing them must be equal with respect to their 'absolute values'; one then has $f = f'$, from which one can infer as a condition of their equilibrium that $f - f' = 0$, which is to say that the algebraic sum of the two forces, or of the two 'vectors' representing

1. Cf. *The Reign of Quantity and the Signs of the Times*, chap. 2.

them, is null, such that equilibrium is thus defined by zero. Zero having been incorrectly regarded by mathematicians as a sort of symbol for nothingness, as we have already said above—as if nothingness could really be symbolized by anything whatsoever—the result seems to be that equilibrium is the state of non-existence, which is a rather strange consequence; it is nevertheless almost certainly for this reason that, instead of saying that two forces in equilibrium neutralize one another, which would be more exact, it is said that they cancel one another, which is contrary to the reality of things, as we have just made clear by a most elementary observation.

The true notion of equilibrium is something else altogether. In order to understand it, it suffices to point out that all natural forces, and not only mechanical forces (which, let us say again, are no more than a very particular case) but forces of the subtle order as well as those of the corporeal order, are either attractive or repulsive; the first can be considered as compressive forces, or forces of contraction, and the second as expansive forces, or forces of dilation,[2] and basically this is no more than an expression in a particular domain of the fundamental cosmic duality itself. It is easy to understand how, given an initially homogenous medium, for every point of compression there will necessarily correspond an equivalent expansion at another point, and inversely, such that two centers of force must be envisaged correlatively, each of which could not exist without the other; this is what one can call the law of polarity, which is, in all its various forms, applicable to all natural phenomena, since it, too, derives from the duality of the very principles that preside over all of manifestation; in the specialized domain with which physicists occupy themselves, this law is above all evident in

---

2. If one considers the ordinary notion of centripetal and centrifugal forces, one will easily see that the first fall under the category of compressive forces, the second under that of expansive forces; likewise, frictional force can be assimilated to the expansive forces, since it is exerted away from its point of application, and an impulse or impact can be assimilated to the compressive forces, since it is on the contrary exerted toward its point of application; but if one envisages things with respect to the point of emission, the inverse will be true, and this is moreover demanded by the law of polarity. In another domain, Hermetic 'coagulation' and 'solution' also correspond to compression and expansion, respectively.

electrical and magnetic phenomena, but it is by no means limited to them. Now if two forces, the one compressive, the other expansive, act upon the same point, then the condition requisite for them to be in equilibrium, or to neutralize one another, the condition, that is, which when fulfilled will produce neither contraction nor dilation, is that the intensities of the two forces be equivalent; we do not say equal, since the forces are of different species, and since this is moreover a question of a truly qualitative, and not simply quantitative, difference. The forces can be characterized by coefficients proportional to the contraction or dilation they produce, in such a way that if one considers a compressive force and an expansive force together, the first will have a coefficient $n > 1$, and the second a coefficient $n' < 1$; each of these coefficients will be the ratio of the density of the space surrounding the point in consideration, under the action of the corresponding force, to the original density of the same space, which in this regard is taken to be homogenous when not subject to any forces in virtue of a simple application of the principle of sufficient reason.[3] When neither compression nor dilation is produced, the ratio is necessarily equal to one, since the density of the space is unchanged; in order for two forces acting upon a point to be in equilibrium, their resultant must have a coefficient of one. It is easy to see that the coefficient of this resultant is the product, and not, as in the ordinary conception, the sum of the coefficients of the two forces under consideration; these two coefficients, n and n', must therefore each be the inverse of the other: $n' = 1/n$, and we will then have $(n)(n') = 1$ as the condition for equilibrium; equilibrium will thus no longer be defined by zero, but by the unit.[4]

It will be seen that the definition of equilibrium with respect to the unit—its only real definition—corresponds to the fact that the unit occupies the mid-point in the doubly indefinite sequence of

3. When we speak thus of the principle of sufficient reason, we of course have in mind only the principle in itself, apart from any of the specialized and more or less contestable forms that Leibniz or others may have wished to give it.

4. This formula corresponds exactly to the conception in Far-Eastern cosmology of the equilibrium of the two complementary principles of *yang* and *yin*.

whole numbers and their inverses, while this central position is as it were usurped by zero in the artificial sequence of positive and negative numbers. Far from being the state of non-existence, equilibrium is on the contrary existence considered in and of itself, independent of its secondary, multiple manifestations; moreover, it is certainly not Non-Being, in the metaphysical sense of the word, for existence, even in this primordial and undifferentiated state, is still the point of departure for all differentiated manifestations, just as the unit is the point of departure for the multiplicity of numbers. As we have just considered it, this unit in which equilibrium resides is what the Far-Eastern tradition calls the 'Invariable Middle'; and according to the same tradition, this equilibrium or harmony is the reflection of the 'Activity of Heaven' at the center of each state, and of each modality of being.

18

Variable and
Fixed Quantities

Let us now return to the question of the justification of the rigor of
the infinitesimal calculus. We have already seen that Leibniz considers quantities to be equal when their difference, while not strictly
null, is nonetheless incomparable with respect to the quantities
themselves; in other words, infinitesimal quantities, though not
nihila absoluta [absolute nothingness], are nevertheless *nihila
respectiva* [nothingness in some respect], and as such must be negligible with respect to ordinary quantities. Unfortunately, the notion
of 'incomparability' is still too imprecise for an argument based on
it alone to be fully sufficient to establish the rigorous character of
the infinitesimal calculus fully; from this point of view, the calculus
appears to be in short but a method of indefinite approximation,
and we cannot say with Leibniz that 'once this is affirmed, it follows
not only that the error is infinitely small, but that it is nothing at
all';[1] but is there no more rigorous means of arriving at this conclusion? We must at least admit that the error introduced into our calculations can be rendered as small as desired, which is already
saying a great deal; but does not precisely this infinitesimal character of the error do away with it completely when one considers, not
only the course of the calculation itself, but its final results?

An infinitesimal difference, that is, one decreasing indefinitely,
can only be the difference between two variable quantities, for it is
obvious that the difference between two fixed quantities can itself
only be a fixed quantity; it would thus be meaningless to speak of an
infinitesimal difference between two fixed quantities. Hence, we

1. Fragment dated from March 26, 1676.

have the right to say that two fixed quantities are 'rigorously equal the moment that their would be difference can be supposed as small as one likes';[2] now, 'the infinitesimal calculus, like ordinary calculation, really has in view only fixed and determined quantities';[3] in short, it introduces variable quantities only as auxiliaries having a purely transitory character, and these variables must disappear from the results, which can only express ratios between fixed quantities. Thus, in order to obtain these results, one must pass from a consideration of variable quantities to one of fixed quantities; and this passage has precisely as its result the elimination of infinitesimal quantities, which are essentially variable, and which can appear only as the differences between variable quantities.

It will now be easy to understand why, in the definition we cited earlier, Carnot insisted that infinitesimal quantities as employed in the calculus, are able to be rendered as small as one likes 'without one's being obliged on that account to vary the quantities to which they are compared.' It is because these latter quantities must in reality be fixed quantities; it is true that in the calculus they are considered to be limits of variable quantities, but these latter merely play the role of simple auxiliaries, as do the infinitesimal quantities which they bring with them. In order to justify the rigor of the infinitesimal calculus, the essential point is that only fixed quantities must figure in the results; in terms of the calculus, therefore, it is ultimately necessary to pass from variable quantities to fixed quantities, and this is indeed a 'passage to the limit', but not as conceived by Leibniz, since there is no result or 'final term' of the variation itself; now—and this is what really matters—the infinitesimal quantities are eliminated of themselves in this passage, and this quite simply by reason of the substitution of fixed quantities for variable quantities.[4]

---

2. Carnot, *Réflexions sur la Métaphysique du Calcul infinitésimal*, p29.

3. Ch. de Freycinet, *De l'Analyse infinitésimale*, preface, pviii.

4. Cf. Ch. de Freycinet, ibid., p220: 'The equations Carnot called "imperfect" are properly speaking unfulfilled equations, or equations of transition, which are rigorous insofar as they are made to serve only for the calculation of limits; they would be absolutely inaccurate, on the contrary, if their limits did not actually have to be found. In order for there to be no doubt as to the value of the ratios through

The Metaphysical Principles of the Infinitesimal Calculus

But must one view their elimination merely as the result of a simple 'compensation of errors', as Carnot would have it? We think not, and it indeed seems that one really can see more in it as soon as one distinguishes between variable and fixed quantities, observing that they constitute as it were two separate domains, between which there doubtless exists a correlation and analogy—which moreover is necessary in order to be able to pass from one to the other, however such a passage is effected—but without their real ratios ever establishing any kind of interpenetration, or even continuity; furthermore, this implies that an essentially qualitative difference exists between the two sorts of quantity, in conformity with what was said earlier concerning the notion of the limit. Leibniz never made this distinction clearly, and here again, his conception of a universally applicable continuity no doubt prevented him from doing so; he was unable to see that 'passage to the limit' essentially implies a discontinuity, because for him no discontinuity existed. However, it is this distinction alone that allows us to formulate the following proposition: if the difference between two variable quantities can be rendered as small as one likes, then the fixed quantities that correspond to these variables and which are regarded as the respective limits of the latter, are rigorously equal. Thus, an infinitesimal difference can never become nothing; but such a difference can exist only between variables, and between the corresponding fixed quantities, the difference must indeed be nothing; whence it immediately follows that to an error capable of being rendered as small as one likes in the domain of variable quantities (in which there can in fact be no question of anything more than indefinite approximation precisely by reason of the character of these quantities) there necessarily corresponds another error that is rigorously null in the domain of fixed quantities. The true justification for the rigor of the infinitesimal calculus essentially resides in this consideration alone, and not in any others, which, whatever they might be, are always more or less peripheral to the question.

---

which one passes, it suffices to keep in mind the actual destination of the calculations. With each of the ratios, one must look not at what it seems to express at the moment, but at that which it will later express, after its limits have been found'.

19

Successive
Differentiations

The preceding still leaves a difficulty regarding the consideration of different orders of infinitesimal quantity: how can one conceive of quantities as infinitesimal not only with respect to ordinary quantities, but with respect to other quantities that are themselves infinitesimal? Here again Leibniz has recourse to the notion of 'incomparables', but this is much too vague to satisfy us, and it does not sufficiently explain the possibility of successive differentiations. No doubt, this possibility can best be understood by a comparison or example from mechanics: 'As for ddx, it is to dx as the *conatus* [force] of weight or the centrifugal tendency is to speed.'[1] And Leibniz develops this idea in his response to the objections of the Dutch mathematician Nieuwentijt, who, while admitting differentials of the first order, maintained that those of higher orders could only be null quantities:

Ordinary quantity, the first infinitesimal or differential quantity, and the second infinitesimal or diffentio-differential quantity, are to each other as movement, speed, and solicitation,[2] which is an element of speed. Movement describes a line, speed an element of the line, and solicitation an element of the element.[3]

1. Letter to Huygens, October 1–11, 1693.
2. By 'solicitation' is meant that which is commonly designated by the term 'acceleration'.
3. *Responsio ad nonnullas difficultates a Dn. Bernardo Nieuwentijt circa Methodum differentialem seu infinitesimalem notas* [The Answer to Several Difficulties Raised by Mr Bernard Nieuwentijt About the Differential or Infinitesimal Method], in the *Acta Eruditorum* of Leipzig, 1695.

But here we have only a particular example or case, which can in short serve only as a simple 'illustration', not an argument, and it is necessary to furnish justification of a general order, which this example, moreover, in a certain sense contains implicitly.

Indeed, differentials of the first order represent the increases—or, better, the variations, since depending on the case they could as easily be in the decreasing as in the increasing direction—that are at each instant received by ordinary quantities; such is speed with respect to the space covered in a given movement. In the same way, differentials of a given order represent the instantaneous variations of differentials of the preceding order, which in turn are taken as magnitudes existing within a certain interval; such is acceleration with respect to speed. Thus the distinction between different orders of infinitesimal quantities in fact rests on the consideration of different degrees of variation, much more than on that of incomparable magnitudes.

In order to state precisely the way in which this must be understood, let us simply make the following remark: one can establish among the variables themselves distinctions analogous to those established earlier between fixed and variable quantities; under these conditions, to go back once again to Carnot's definition, a quantity is said to be infinitesimal with respect to others when one can render it as small as one likes 'without one being obliged thereby to vary these other quantities.' Indeed, this is because a quantity that is not absolutely fixed, or even one that is essentially variable—as is the case with infinitesimal quantities, whatever the order in question—can nevertheless be regarded as fixed and determined, that is, as capable of playing the role of fixed quantity with respect to certain other variables. Only under these conditions can a variable quantity be considered the limit of another variable, which, by the very definition of the term limit, presupposes that it be regarded as fixed, at least in a certain respect, namely relative to that which it limits; inversely, a quantity can be variable not only in and of itself or, what amounts to the same, with respect to absolutely fixed quantities, but even with respect to other variables, insofar as the latter are regarded as relatively fixed.

Instead of speaking in this regard of degrees of variation, as we have just done, one could equally well speak of degrees of indeterminacy, which ultimately would be exactly the same thing, only considered from a slightly different point of view: a quantity, though indeterminate by its nature, can nevertheless be determined in a relative sense by the introduction of certain hypotheses, which allow the indeterminacy of other quantities to subsist at the same time; these latter quantities will therefore be more indeterminate, so to speak, than the others, or indeterminate to a greater degree, and they will therefore be related to the others in a manner comparable to that in which the indeterminate quantities are themselves related to quantities that truly are determined. We shall confine ourselves to these remarks on the subject, for however summary they might be, we believe that they are at least sufficient for understanding the possibility of the existence of differentials of various successive orders; but, in connection with this same question, it still remains for us to show more explicitly that there is really no logical difficulty in considering multiple degrees of indefinitude, and this as much in the order of decreasing quantities, to which infinitesimals and differentials belong, as in that of increasing quantities, in which one can likewise envisage integrals of different orders, which are as it were symmetric with respect to the successive differentiations; and this is moreover in conformity with the correlation that exists between the indefinitely increasing and the indefinitely decreasing, as we have explained. Of course, in all this it is only a question of degrees of indefinitude, and not of 'degrees of infinity', such as Jean Bernoulli understood them, which notion Leibniz dared neither adopt nor reject absolutely in this regard; and here we have yet another case in which the difficulties can be immediately resolved by substituting the notion of the indefinite for that of the so-called infinite.

20

Various Orders
of Indefinitude

The logical difficulties, and even contradictions which mathematicians run up against when they consider 'infinitely great' or 'infinitely small' quantities that differ with respect to one another, and even belong to different orders altogether, arise solely from the fact that they regard as infinite that which is simply indefinite. It is true that in general they do not seem very concerned with these difficulties, but they exist nonetheless, and are no less serious for all that, as they cause the science of mathematics to appear as if full of illogicalities, or, if one prefer, of 'para-logicalities', and such a science loses all real value and significance in the eyes of those who do not allow themselves to be deluded by words. Here are some examples of the contradictions introduced by those who would allow the existence of infinite magnitudes, when they apply this notion to geometric magnitudes: if a straight line is considered to be infinite, its infinitude must be less, and even infinitely less, than the infinitude constituted by a surface such as a plane, in which both that line and an infinite number of others are also contained, and the infinitude of the plane will in turn be infinitely less than that of three-dimensional space. The very possibility of the coexistence of all of these would-be infinities, some of which are supposed to be infinite to the same degree, others to different degrees, suffices to prove that none of them can be truly infinite, even apart from any consideration of a more properly metaphysical order; indeed, as these are truths which we cannot emphasize enough, let it be said again: it is obvious that if one supposes a plurality of distinct infinites, each will have to be limited by the others, which amounts to saying that they will exclude one another. Moreover, to tell the truth, the 'infinitists', for whom this purely verbal accumulation of an 'infinity of infinities' seems to produce a kind of 'mental intoxication', if such an expression be permissible, do not retreat in face of such contradictions, since, as has already been said, they see no difficulty in asserting that various infinite numbers exist, and that consequently one infinity can be greater or smaller than another; but the absurdity of such utterances is only too obvious, and the fact that they are commonly used in contemporary mathematics changes nothing, but only shows to what extent the sense of the most elementary logic has been lost in our day. Yet another contradiction, no less blatant than the last, is to be found in the case of a closed, hence obviously and visibly finite, surface, which nevertheless contains an infinite number of lines, as, for example, a sphere, which contains an infinite number of circles; here we have a finite container, of which the contents would be infinite, which is likewise the case, moreover, when one maintains, as did Leibniz, the 'actual infinity' of the elements of a continuous set.

On the contrary, there is no contradiction in allowing the coexistence of a multiplicity of indefinite magnitudes of various orders. Thus a line indefinite in a single dimension can in this regard be considered to constitute a simple indefinitude of the first order; a surface, indefinite in two dimensions, and embracing an indefinite number of indefinite lines, will then be an indefinitude of the second order; and three-dimensional space, which embraces an indefinite number of indefinite surfaces, will similarly be an indefinitude of the third order. Here it is essential to point out once again that we said the surface embraces an indefinite number of lines, not that it is constituted by an indefinite number of lines, just as a line is not composed of points, but rather embraces an indefinite multitude of them; and it is again the same in the case of a volume with respect to its surfaces, three-dimensional space being itself none other than an indefinite volume. This, moreover, is basically what we said above on the subject of 'indivisibles' and the 'composition of the continuous'; it is questions of this kind that, precisely by reason of their complexity, most make one aware of the necessity of rigorous language. Let us also add in this regard that if from a certain point of view one can legitimately consider a line to be generated by a point, a surface by a line, and a volume by a surface, this essentially presupposes that the point, the line, or the surface be displaced through a continuous motion, embracing an indefinitude of successive positions; and this is altogether different from considering each of these positions in isolation, that is, regarding the points, lines, and surfaces as fixed and determined, and as constituting the parts or elements of the line, the surface, or the volume, respectively. Likewise, but inversely, when one considers a surface to be the intersection of two volumes, a line the intersection of two surfaces, and a point the intersection of two lines, these intersections must not, of course, by any means be conceived of as parts common to the volumes, surfaces, or lines; they are only limits or extremities of the latter, as Leibniz has said.

According to what we have just said, each dimension introduces as it were a new degree of indeterminacy to space, that is, to the spatial continuum insofar as it is subject to indefinite increase of extension and thus yields what could be called successive powers of the indefinite;[1] and one can also say that an indefinite quantity of a certain order or power contains an indefinite multitude of indefinite quantities of a lower order or lesser power. As long as it is only a question of the indefinite in all of this, these considerations, as well as others of the same sort, remain perfectly acceptable, for there is no logical incompatibility between multiple and distinct indefinite quantities, which, despite their indefinitude, are nonetheless of an essentially finite nature, and which, like any other particular and determined possibility, are therefore perfectly capable of coexisting within total Possibility, which is alone infinite, since it is identical to the universal All.[2] These same considerations take on an impossible and absurd form only when the indefinite is confused with the infinite; thus, as with the notion of the 'infinite multitude', we once again have an instance in which the contradiction inherent in a so-called determined infinite is concealed, deforming another idea that, although in itself not at all contradictory, is nonetheless rendered virtually unrecognizable.

1. Cf. *The Symbolism of the Cross*, chap. 12.
2. Cf. *The Multiple States of the Being*, chap. 1.

We have just spoken of various degrees of indeterminacy in relation to quantities taken in the increasing direction; by applying the same notion to the decreasing direction we have already justified above the consideration of various orders of infinitesimal quantity, the possibility of which is all the more understandable in the light of the correlation we noted earlier between indefinitely increasing and indefinitely decreasing quantities. Among indefinite quantities of various orders, those of orders apart from the first will always be indefinite with respect to those of the preceding order as well as to ordinary quantities; inversely, among infinitesimal quantities of various orders, it is just as legitimate to consider those of each order as infinitesimal not only with respect to ordinary quantities, but also to the infinitesimal quantities of the preceding orders.[3] There is no absolute heterogeneity between indefinite quantities and ordinary quantities, nor again between infinitesimal quantities and ordinary quantities; in short, it is only a question of a difference of degree, not of kind, since, in reality, the consideration of indefin-itude, whatever the order or power in question, never takes us out of the finite; again, it is the false conception of the infinite that introduces the appearance of a radical heterogeneity between the different orders of quantity, which at bottom is completely incomprehensible. In doing away with this heterogeneity, a kind of continuity is established quite different from that which Leibniz envisaged between variables and their limits, and much better grounded in reality, for contrary to what he believed, the distinction between variable and fixed quantities essentially implies a difference of nature.

3. In accordance with common usage, we reserve the denomination 'infinitesimal' for indefinitely decreasing quantities, to the exclusion of indefinitely increasing quantities, which, for the sake of convenience, we can call simply 'indefinite'; it is rather strange that Carnot brought both together under the name of 'infinitesimal', contrary not only to common usage but even to the obvious origin of the term. While we shall continue to use the word 'infinitesimal' in the sense just given, we cannot refrain from pointing out that the term has one serious shortcoming, namely that it is clearly derived from the word 'infinite', which renders it scarcely adequate to the idea it really expresses; to be able to use it without any drawbacks, its origin must be forgotten, so to speak, or at least accorded a solely 'historical' character, as arising from Leibniz's conception of 'well-founded fictions'.

The Metaphysical Principles of the Infinitesimal Calculus

Under these conditions, ordinary quantities themselves can in a way be regarded as infinitesimal with respect to indefinitely increasing quantities, at least when we are dealing with variables, for, if a quantity is capable of being rendered as great as one likes with respect to another, inversely the latter will by the same token become as small as one likes with respect to the former. We say that it must be a question of variables because an infinitesimal quantity must always be conceived of as essentially variable, and this restriction is inherent in its very nature; moreover, quantities belonging to two different orders of indefinitude are inevitably variable with respect to one another, and this property of relative and reciprocal variability is perfectly symmetric, for, in accordance with what was just said, to consider one quantity to be indefinitely increasing with respect to another, or this latter indefinitely decreasing with respect to the first, amounts to the same thing; without this relative variability there could be neither indefinite increase nor indefinite decrease, but only definite and determined ratios between the two quantities.

In the same way, whenever there is a change in position with respect to two bodies A and B, to say that body A is in motion with respect to body B, and, inversely, that body B is in motion with respect to body A, also amounts to the same thing, at least insofar as the change is only considered in and of itself; in this regard the concept of relative motion is just as symmetric as that of relative variability, which we were just considering. This is why, according to Leibniz, who used it to demonstrate the inadequacy of Cartesian mechanism as a physical theory claiming to furnish an explanation for all natural phenomena, one cannot distinguish between a state of motion and a state of rest when one is limited solely to the consideration of changes in position; to do so one must bring in something of another order, namely, the notion of force, which is the proximate cause of such changes, and which alone can be attributed to one body rather than to another, as it allows the true cause of change to be located in one body and in that body alone.[4]

---

4. See Leibniz, *Discours de Métaphysique*, chap. 18; cf. *The Reign of Quantity and the Signs of the Times*, chap. 14.

21

The Indefinite
is Analytically
Inexhaustible

In the two cases just considered, that of the indefinitely increasing and that of the indefinitely decreasing, a quantity of a given order can be regarded as the sum of an indefinitude of elements, each of which is an infinitesimal quantity with respect to the entire sum. In order to be able to speak of infinitesimal quantities, it is moreover necessary that it be a question of elements that are not determined with respect to their sum, and this is indeed the case whenever the sum is indefinite with respect to the elements in question; this follows immediately from the essential character of indefinitude itself, inasmuch as the latter obviously implies the idea of 'becoming', as we have said before, and consequently a certain indeterminacy. It is of course understood that this indeterminacy can only be relative, and exists only from a certain point of view or with respect to a certain thing: such is the case, for example, with a sum that is an ordinary quantity, and hence not indefinite in and of itself, but only with respect to its infinitesimal elements; at any rate, if it were otherwise, and if this notion of indeterminacy were not introduced, one would be reduced to the mere conception of 'incomparables', interpreted in the crude sense of the grain of sand in comparison to the earth, and the earth in comparison to the heavens.

The sum in question can by no means be effected in the manner of an arithmetical sum, since for that it would be necessary for an indefinite series of successive additions to be achieved, which is contradictory; in the case in which the sum is an ordinary and determined quantity as such, it is obviously necessary, as we already said when we set forth the definition of the integral calculus, that the number, or rather the multitude, of elements increase indefinitely while at the same time the magnitude of each decreases indefinitely, and in this sense the indefinitude of its elements is truly inexhaustible. But if the sum cannot be effected in this way, as the final result of a multitude of distinct and successive operations, it can on the other hand be comprehended at one stroke, by a single operation, namely, integration;[1] here we have the inverse operation of differentiation, since it reconstitutes the sum starting from its infinitesimal elements, while differentiation on the contrary moves from the sum to the elements, furnishing the means of formulating the law for the instantaneous variations of the quantity of which the expression is given.

Thus, whenever it is a question of indefinitude, the notion of an arithmetical sum is no longer applicable, and one must resort to the notion of integration in order to compensate for the impossibility of 'numbering' the infinitesimal elements, an impossibility which, of course, results from the very nature of these elements, and not from any imperfection on our part. In passing we may observe that as regards the application of this to geometric magnitudes (which, moreover, is ultimately the true raison d'être of the infinitesimal calculus), this is a method of measurement completely different from the usual method founded on the division of a magnitude into definite portions, of which we spoke previously in connection with 'units of measurement'. The latter always amounts in short to a substitution of the discontinuous for the continuous by 'cutting up' the sum into various portions equal to a magnitude of the same species

---

1. The terms 'integral' and 'integration', which have prevailed in usage, are not Leibniz's, but Jean Bernoulli's; in their place Leibniz used only the words 'sum' and 'summation', with the drawback that these terms seem to indicate an analogy between the operation in question and the formation of an arithmetical sum; we say only that they seem to do so, for it is quite certain that the essential difference between the two operations could not have escaped Leibniz.

taken as the unit,[2] in order to be able to apply the resulting number directly to the measurement of continuous magnitudes, which cannot actually be done except by altering the nature of the magnitudes in order to make it assimilable, so to speak, to that of number. The other method, on the contrary, respects the true character of continuity as much as possible, regarding it as a sum of elements that are fixed and determined, but that are essentially variable and by virtue of their variability capable of becoming smaller than any assignable magnitude; this method thereby allows the spatial quantity between the limits of these elements to be reduced as much as one likes, and it is therefore the least imperfect representation of continuous variation one can give, in that it takes account of the nature of number, which in spite of everything cannot be changed.

These observations will allow us to understand more precisely in what sense one can say, as we did at the beginning, that the limits of the indefinite can never be reached through any analytical procedure, or, in other words, that the indefinite, while not absolutely and in every way inexhaustible, is at least analytically inexhaustible. In this regard, we must naturally consider those procedures analytical which, in order to reconstitute a whole, consist in taking its elements distinctly and successively; such is the procedure for the formation of an arithmetical sum, and it is precisely in this regard that it differs essentially from integration. This is particularly interesting from our point of view, for one can see in it, as a very clear example, the true relationship between analysis and synthesis: contrary to current opinion, according to which analysis is as it were a preparation for synthesis, or again something leading to it, so much so that one must always begin with analysis, even when one does not intend to stop there, the truth is that one can never actually arrive at synthesis through analysis. All synthesis, in the true sense of the word, is something immediate, so to speak, something that is

---

2. Or by a fraction of this magnitude, which matters little, since the fraction would then constitute a secondary, smaller unit substituted for the first in the case in which division by the original magnitude cannot be carried out exactly; and, in order to obtain an exact, or least a more exact, result, one instead uses this fraction.

not preceded by any analysis and is entirely independent of it, just as integration is an operation carried out in a single stroke, by no means presupposing the consideration of elements comparable to those of an arithmetical sum; and as this arithmetical sum can yield no means of attaining and exhausting the indefinite, this latter must, in every domain, be one of those things that by their very nature resist analysis and can be known only through synthesis.[3]

3. Here, and in what follows, it should be understood that we take the terms 'analysis' and 'synthesis' in their true and original sense, and one must indeed take care to distinguish this sense from the completely different and quite improper sense in which one currently speaks of 'mathematical analysis', according to which integration itself, despite its essentially synthetic character, is regarded as playing a part in what one calls 'infinitesimal analysis'; it is for this reason, moreover, that we prefer to avoid using this last expression, availing ourselves only of those of 'the infinitesimal calculus' and 'the infinitesimal method', which lead to no such equivocation.

22

The Synthetic
Character of Integration

Contrary to the formation of an arithmetical sum, which, as we have just said, is strictly analytic in character, integration must be regarded as an essentially synthetic operation in that it simultaneously embraces each element of the sum to be calculated, preserving the 'indistinction' appropriate to the parts of a continuum, since, by the very nature of continuity, these parts cannot be fixed and determined things. Moreover, whenever one wishes to calculate the sum of the discontinuous elements of an indefinite sequence, this 'indistinction' must likewise be maintained, although for a slightly different reason, for even if the magnitude of each may be conceived of as determined, the total number of elements may not, and we can even say more exactly that their multitude surpasses all number; nevertheless, there are some cases in which the sum of the elements of such a sequence tends toward a certain definite limit, even when their multitude increases indefinitely. Although such a manner of speaking might at first seem a little strange, one could also say that such a discontinuous sequence is indefinite by 'extrapolation', while a continuous set is so by 'interpolation'; what is meant by this is that if one takes a given portion of a discontinuous sequence, bounded by any two of its terms, such a portion will in no way be indefinite, as it is determined both as a whole and with respect to its elements; the indefinitude of the sequence lies in the fact that it extends beyond this portion, without ever arriving at a final term; on the contrary, the indefinitude of a continuous set, determined as such, is to be found precisely in its interior, since its elements are not determined, and since it has no final terms, the continuous being always divisible; in this respect each case is thus as it were the inverse of the other. The summation of an indefinite numerical sequence will never be completed if each term must be taken one by one, since there is no final term whereby the sequence could come to an end; such a summation is possible only in the case where a synthetic procedure lets us seize in a single stroke, as it were, the indefinitude considered in its entirety, without this at all presupposing the distinct consideration of its elements, which, moreover, is impossible, by the very fact that they constitute an indefinite multitude. And similarly, when an indefinite sequence is given to us implicitly by its law of formation, as in the case of the sequence of whole numbers, we can say that it is thus given to us completely in a synthetic manner, and that it cannot be given otherwise; indeed, to do so analytically would be to lay out each term distinctly, which is an impossibility.

Therefore, whenever we have a given example of indefinitude to consider, whether it be a continuous set or a discontinuous sequence, it will be necessary in every case to have recourse to a synthetic operation in order to reach its limits; progression by degrees would be useless here and could never bring us to our goal, for such a progression can arrive at a final term only on the twofold condition that both this term and the number of degrees to be covered in order to reach it, be determined. That is why we did not say that the limits of the indefinite could not be reached at all, which would be unjustifiable when its limits do exist, but only that they cannot be reached analytically: the indefinite cannot be exhausted by degrees, but it can be embraced in its totality by certain transcendent operations, of which integration is the classic example in the mathematical order. One could point out that progression by degrees here corresponds precisely to the variation of quantity, directly in the case of discontinuous sequences and, in cases of continuous variation, following therefrom, so to speak, to the extent permitted by the discontinuous nature of number; on the other hand, synthetic operations immediately place one outside of and beyond the domain of variation, as must necessarily be the case according with what we said above, in order for a 'passage to the limit' actually to be realized; in other words, analysis pertains only to variables, taken in the very course of their variation, while synthesis alone attains their limits, which is the only definitive and really valuable result, since, to be able to speak of results, one must clearly arrive at something relating exclusively to fixed and determined quantities.

Furthermore, one can of course find analogous synthetic operations in domains apart from quantity, for the idea of an indefinite development of possibilities is clearly applicable to other things than quantity, as, for example, to a given state of manifested existence and the conditions, whatever they might be, to which the state is subject, whether considered with respect to the whole of the cosmos, or to one being in particular; that is, one can take either a 'macrocosmic' or a 'microcosmic' point of view.[1] One could say that in this case 'passage to the limit' corresponds to the definitive fixation of the results of manifestation in the principial order; indeed, by this alone does the being finally escape from the change and 'becoming' that is necessarily inherent to all manifestation as such; and one can thus see that this fixation is in no way a 'final term' of the development of manifestation, but rather that it is essentially situated outside of and beyond that development, since it belongs to another order of reality, transcendent in relation to manifestation and 'becoming'; in this regard, the distinction between the manifested order and the principial order thus corresponds analogically to that which we established between the domains of variable and fixed quantities. What is more, when it is a question of fixed quantities, it is obvious that no modification can be introduced by any operation whatsoever, and that, consequently, 'passage to the limit' cannot produce anything in this domain, but can only give us knowledge of it; likewise, the principial order being immutable, arriving at it is not a question of 'effectuating' something that did not exist before, but rather of effectively taking cognizance, in a permanent and absolute manner, of that which is. Given the subject of this study, we must naturally consider more particularly and above all, what properly concerns the quantitative domain, in which, as we have seen, the idea of the development of possibilities is trans-

---

1. On this analogical application of the notion of integration, cf. *The Symbolism of the Cross*, chaps. 18 and 20.

lated by the notion of variation, whether in the direction of indefinite increase or of indefinite decrease; but these few will suffice to show that by an appropriate analogical transposition all of this is capable of receiving an incomparably greater significance than that which it appears to have in and of itself, since integration and other operations of the same kind will thereby veritably appear as symbols of metaphysical 'realization' itself.

By this one sees the extent of the difference between traditional science, which allows such considerations, and the profane science of the moderns; and, in this connection, we shall add yet another remark directly relating to the distinction between analytic and synthetic knowledge. Profane science, indeed, is essentially and exclusively analytical; it never considers principles, losing itself instead in the details of phenomena, of which the indefinite and indefinitely changing multiplicity are for it truly inexhaustible, such that it can never arrive at any real or definitive result as far as knowledge is concerned; it keeps solely to phenomena themselves, that is, to exterior appearances, and is incapable of reaching the heart of things, for which Leibniz had already reproached Cartesian mechanism. This is moreover one of the reasons by which modern 'agnosticism' is explained, for, since there are things that can be known only synthetically, whoever proceeds by analysis alone is thereby led to declare such things 'unknowable', since in this respect they really are so, just as those who keep to the analytic view of the indefinite believe its indefinitude to be absolutely inexhaustible, whereas in reality it is so only analytically. It is true that synthetic knowledge is essentially what one might call 'global' knowledge, as is the knowledge of a continuous set or an indefinite sequence the elements of which are not and cannot be set out distinctly; but, apart from the fact that this knowledge is ultimately all that really matters, one can always—since everything is contained in it in principle—descend from it to the consideration of such particular things as one might wish, just as, if an indefinite sequence, for example, is given synthetically through the knowledge of its law of formation, one can as occasion arises always calculate any of its particular terms, while on the contrary when one takes as one's starting-point these same particular things considered in and of themselves, and in all their indefinite detail, one can never rise to the level of principles; and, as we said at the beginning, it is in this regard that the method and point of view of traditional science is as it were inverse to that of profane science, as synthesis itself is to analysis. Moreover, we have here only an application of the obvious truth that, although the 'lesser' can be drawn from the 'greater', one can never cause the 'greater' to come from the 'lesser'; nevertheless, this is precisely what modern science claims to do, with its mechanistic and materialistic conceptions and its exclusively quantitative point of view; but it is precisely because this is impossible that such science is, in reality, incapable of giving the true explanation of anything whatever.[2]

2. On this last point, one can again refer to the considerations set forth in *The Reign of Quantity and the Signs of the Times.*

23

The Arguments of
Zeno of Elea

The preceding considerations implicitly contain the solution to all problems of the sort raised by Zeno of Elea in his famous arguments against the possibility of motion, or at least in what appear to be such when one takes the arguments only as they are usually presented; in fact, one might well doubt whether this was really their true significance. Indeed, it is rather unlikely that Zeno really intended to deny motion; what is more probable is that he merely wished to prove the incompatibility of the latter with the supposition, accepted notably by the atomists, of a real, irreducible multiplicity existing in the nature of things. It was therefore originally against this very multiplicity so conceived that these arguments originally must have been directed; we do not say against all multiplicity, for it goes without saying that multiplicity also exists within its order, as does motion, which, moreover, like every kind of change, necessarily supposes multiplicity. But just as motion, by reason of its character of transitory and momentary modification, is not self-sufficient and would be purely illusory were it not linked to a higher principle transcendent with respect to it, such as the 'unmoved mover' of Aristotle, so multiplicity would truly be non-existent were it to be reduced to itself alone, and did it not proceed from unity, as is reflected mathematically in the formation of the sequence of numbers, as we have seen. What is more, the supposition of an irreducible multiplicity inevitably excludes all real connections between the elements of things, and consequently all continuity as well, for the latter is only a particular case or special form of such connections. As we have already said above, atomism necessarily implies the discontinuity of all things; ultimately, motion really is incompatible with this discontinuity, and we shall see that this is indeed what the arguments of Zeno show.

Take, for example, the following argument: an object in motion can never pass from one position to another, since between the two there is always an infinity of other positions, however close, that must be successively traversed in the course of the motion, and, however much time is employed to traverse them, this infinity can never be exhausted. Assuredly, this is not a question of an infinity, as is usually said, for such would have no real meaning; but it is no less the case that in every interval one may take into account an indefinite number of positions for the moving object, and these cannot be exhausted in analytic fashion, which would involve each position being occupied one by one, as the terms of a discontinuous sequence are taken one by one. But it is this very conception of motion that is in error, for it amounts in short to regarding the continuous as if it were composed of points, or of final, indivisible elements, like the notion according to which bodies are composed of atoms; and this would amount to saying that in reality there is no continuity, for whether it is a question of points or atoms, these final elements can only be discontinuous; furthermore, it is true that without continuity there would be no possible motion, and this is all that the argument actually proves. The same goes for the argument of the arrow that flies and is nonetheless immobile, since at each instant one sees only a single position, which amounts to supposing that each position can in itself be regarded as fixed and determined, and that the successive positions thus form a sort of discontinuous series. It is further necessary to observe that it is not in fact true that a moving object is ever viewed as if it occupied a fixed position, and that quite to the contrary, when the motion is fast enough, one will no longer see the moving object distinctly, but only the path of its continuous displacement; thus for example, if a flaming ember is whirled about rapidly, one will no longer see the form of the ember, but only a circle of fire; moreover, whether one explains this by the persistence of retinal impressions, as physiologists do, or in any other way, it matters little, for it is no less obvious that in such cases one grasps the continuity of motion directly, as it were, and in a perceptible manner. What is more, when one uses the expression 'at each instant' in formulating such arguments, one is implying that time is formed from a sequence of indivisible instants, to each of which there corresponds a determined position of the object; but in reality, temporal continuity is no more composed of instants than spatial continuity is of points, and as we have already pointed out, the possibility of motion presupposes the union, or rather the combination, of both temporal and spatial continuity.

It is also argued that in order to traverse a given distance, it is first necessary to traverse half this distance, then half of the remaining half, then half of the rest, and so on indefinitely,[1] such that one would always be faced with an indefinitude that, envisaged in this way, is indeed inexhaustible. Another almost equivalent argument is as follows: if one supposes two moving objects to be separated by a certain distance, then one of them, even if traveling faster than the other, will never be able to overtake the other, for, when it arrives at the point where it would have met the one in the lead, the latter will be in a second position, separated from the first by a smaller distance than the initial one; when it arrives at this new position, the other will be in yet a third position, separated from the second by a still smaller distance, and so on indefinitely, in such a way that, despite the fact that the distance between the two objects is always decreasing, it will never disappear altogether. The essential problem with these two arguments, as well as with the preceding, consists in the fact that they all suppose that in order to reach a certain endpoint, all the intermediate degrees must be traversed distinctly and successively. Now, we are led to one of two conclusions: either the motion in question is indeed continuous, and therefore cannot be broken down in this way, since the continuous has no irreducible elements; or the motion is composed, or at least may be considered to be composed, of a discontinuous succession of intervals, each with a determined magnitude, as with the steps taken by a man

---

1. This corresponds to the successive terms of the indefinite series $1/1 + 1/2 + 1/4 + 1/8 + \ldots = 2$, used by Leibniz as an example in a passage already cited above.

walking,[2] in which case the consideration of these intervals would obviously rule out that of all the various intermediate positions possible, which would not actually have to be traversed as so many distinct steps. Besides, in the first case, which is really that of a continuous variation, the end-point, assumed by definition to be fixed, cannot be reached within the variation itself, and the fact that it actually is reached demands the introduction of a qualitative heterogeneity, which this time does constitute a true discontinuity, and which is represented here by the passage from the state of motion to that of rest; this brings us to the question of 'passage to the limit', the true meaning of which still remains to be explained.

2. In reality, the motions comprising his walking are indeed continuous, like any other motion, but the points where he touches the ground form a discontinuous sequence, such that each step marks a determined interval, and the distance traversed can thus be broken down into such intervals, the ground not being touched at any intermediate points.

24

The True Conception
of Passage to the Limit

The consideration of 'passage to the limit', we said above, is necessary, if not to the practical applications of the infinitesimal method,
then at least to its theoretical justification, and this justification is
precisely the only thing that concerns us here, for simple practical
rules of calculation that succeed in an as it were 'empirical' manner
and without our knowing exactly why, are obviously of no interest
from our point of view. Undoubtedly, in order to perform the calculations, and even to follow them through to the end, there is in
fact no need to raise the question as to whether the variable reaches
its limit, or how it can do so; nevertheless, if it does not reach its
limit, such a calculus will only have value as a simple calculus of
approximation. It is true that here we are dealing with an indefinite
approximation, since the very nature of infinitesimal quantities
allows the error to be rendered as small as one might wish, without
it being possible to eliminate it entirely, since despite the indefinite
decrease, these same infinitesimal quantities never become nothing.
Perhaps one might say that, practically speaking, this is the equivalent of a perfectly rigorous calculation; but, besides the fact that this
is not what matters to us, such is not in question, can the indefinite
approximation itself retain meaning if, with respect to the desired
results, one no longer envisages variables, but rather fixed and
determined quantities? Under these conditions, one cannot escape
the following alternative as far as the results are concerned: either
the limit is not reached, in which case the infinitesimal calculus is
then only the least crude of various methods of approximation; or
the limit is reached, in which case one is dealing with a method that

is truly rigorous. But we have seen that limits, by their very definition, can never exactly be reached by variables; how, then, do we
have the right to say that they are nonetheless reached? This can be
precisely accomplished, not in the course of the calculation, but in
the results, since only fixed and determined quantities, like the limit
itself, must figure therein, while variables no longer do so; consequently the distinction between variable and fixed quantities, which
is a strictly qualitative distinction, moreover, is the only true justification for the rigor of the infinitesimal calculus, as we have already
said.

Thus, let us repeat it again, a limit cannot be reached within a
variation, and as a term of the latter; it is not the final value the variable takes on, and the idea of a continuous variation arriving at any
'final value', or 'final state', would be as incomprehensible and contradictory as that of an indefinite sequence arriving at a 'final term',
or of the division of a continuum arriving at 'final elements'. Therefore a limit does not belong to the sequence of successive values of
the variable, but it falls outside of this series, and that is why we said
that 'passage to the limit' essentially implies a discontinuity. Were it
otherwise, we would be faced with an indefinitude that could be
exhausted analytically, and this can never happen. Here the distinction we previously established in this regard takes on its full significance, for we find ourselves in one of those cases in which it is a
question of reaching the limits of a given indefinite quantity,
according to an expression we have already used; it is therefore not
without reason that the same word 'limit' comes up again, but with
another, more specialized meaning, in the particular case we shall
now consider. The limit of a variable must truly limit, in the general
sense of the word, the indefinitude of the states or possible modifi-
cations comprised within the definition of this variable; and it is
precisely for this reason that it must necessarily be located outside
of that which it limits. There can be no question of exhausting this
indefinitude through the very course of the variation by which it is
constituted; in reality, it is a question of passing beyond the domain
of this variation, in which the limit is not contained, and this is the
result that is obtained, not analytically and by degrees, but synthetically and in a single stroke, in a manner that is as it were 'sudden'

and corresponds to the discontinuity produced in passing from variable to fixed quantities.[1]

Limits pertain essentially to the domain of fixed quantities; this is why 'passage to the limit' logically demands the simultaneous consideration of two different and as it were superimposed modalities existing within quantity; it is nothing other than passage to the higher modality, in which what exists only as the state of a simple tendency in the lower modality, is fully realized; to use the Aristotelian terminology, it is a passage from potentiality to actuality, which assuredly has nothing in common with the simple 'compensation of errors' that Carnot had in mind. The mathematical notion of the limit implies by its very definition a character of stability and equilibrium, which applies to permanent and definite things, and which obviously cannot be realized by quantities insofar as one considers them in the lower of the two modalities, as essentially variable; the limit can therefore never be reached gradually, but only immediately by the passage from one modality to the other, which alone allows the omission of all intermediate stages, since it includes and embraces synthetically all of their indefinitude; in this way, what was and could only be but a tendency within the variable, is affirmed and fixed in a real and definite result. Otherwise, 'passage to the limit' would always be an illogicality pure and simple, for it is obvious that, insofar as one keeps to the domain of variables, one cannot obtain the fixity appropriate to limits, since the quantity previously considered to be variable would precisely have to lose its transitory and contingent character. The state of variable quantities is indeed an eminently transitory and as it were imperfect state, since it is only the expression of a 'becoming', as we have likewise found to be the case with the idea at the root of indefinitude itself, which, moreover, is closely linked to the state of variation. The calculation will thus only be perfect, or truly completed, when it arrives at results in which there is no longer anything variable or

1. This 'sudden' or 'instantaneous' character could be compared, by way of an analogy from the order of natural phenomena, to the example we gave above concerning the breaking of the rope: the rupture itself is also a limit, namely of the tension, but it is by no means comparable to tension, whatever the degree.

indefinite, but only fixed and determined quantities; and we have already seen how this can be applied through analogical transposition beyond the quantitative order—which latter will then have no more than a symbolic value—and will extend even to that which directly concerns the metaphysical 'realization' of being.

25

Conclusion

There is no need to stress the importance that the issues examined in the course of this study present from the strictly mathematical point of view, as they contain the solution to all the problems that have been raised concerning the infinitesimal method, whether regarding its true significance or its rigor. The necessary and sufficient condition for arriving at this solution is nothing other than the strict application of true principles, but these are precisely the principles of which modern mathematicians, along with all other profane scholars, are completely ignorant. Ultimately this ignorance is the sole reason for so many of the discussions that, under these conditions, can be pursued indefinitely without ever reaching any valid conclusion, but on the contrary only further confuse the question and multiply the confusions, as the quarrel between the 'finitists' and 'infinitists' shows only too well. Nevertheless all such discussions would have been cut short quite easily had the true notion of the metaphysical Infinite and the fundamental distinction between the Infinite and the indefinite been set forth clearly and before all else. On this subject Leibniz himself, who unlike those who have come after him at least had the merit of frankly facing certain questions, too often says things that are hardly metaphysical, and are sometimes even as clearly anti-metaphysical, as are the ordinary speculations of most modern philosophers; thus it is again this same lack of principles that prevented him from responding to his adversaries in a satisfying and as it were definitive way, and which consequently opened the door to all subsequent discussions. No doubt one can say with Carnot that, 'if Leibniz was mistaken, it was solely in raising doubts as to the exactitude of his own analysis, so far as he really had these doubts';[1] but even if ultimately he did not, he was nonetheless unable to demonstrate its exactitude rigorously since his conception of continuity, which is most certainly neither metaphysical nor logical, prevented him from making the necessary distinctions and consequently from formulating a precise notion of the limit, which is as we have shown of chief importance for the foundation of the infinitesimal method.

From all of this one can see what significance the consideration of principles can have even for a specialized science considered in and of itself, and without any intention of going further in support of this science than the relative and contingent domain to which the principles are immediately applicable. Of course, this is what the moderns totally misunderstand, readily boasting as they do that with their profane conception of science they have rendered the latter independent of metaphysics, and likewise of theology,[2] while the truth of the matter is that they have thereby only deprived it of all real value as far as knowledge is concerned. In addition, once one understands the need to link science back to principles, it goes without saying that there should no longer be any reason to stop there, and one will quite naturally be led back to the traditional conception according to which a particular science, whatever it might be, is less valuable for what it is in itself than for the possibility of using it as a 'support' for elevating oneself to knowledge of a higher order.[3] Our intention here has been to present by way of a characteristic example an idea of precisely what it would be possible to do, at least in certain cases, to restore to science, mutilated and distorted by profane conceptions, its real value and scope, both from the point of view of the relative knowledge it represents directly, and from that of the higher knowledge to which it can lead through analogical transposition. In this last respect we have been able to see, notably,

1. *Réflexions sur la Métaphysique du Calcul infinitésimal*, p33.
2. We recall somewhere having seen a contemporary 'scientist' who was indignant at the fact that in the Middle Ages, for example, the Trinity had been spoken of in connection with the geometry of the triangle; he probably did not suspect that this is still the case today in the symbolism of the 'Compagnonnage'.
3. For an example on this subject, see *The Esoterism of Dante*, chap. 2, on the esoteric or initiatic aspect of the 'liberal arts' of the Middle Ages.

what may be drawn from notions such as those of integration and 'passage to the limit'. Moreover, it should be said that, more than any other science, mathematics thus furnishes a particularly apt symbolism for the expression of metaphysical truths to the extent that the latter are expressible, as those familiar with some of our other works are aware. This is why mathematical symbolism is used so frequently, whether from the traditional point of view in general, or from the initiatic point of view in particular.[4] But it is of course understood that in order for this to be so it is above all necessary that these sciences be rid of the various errors and confusions that have been introduced by the false views of the moderns, and we should be happy if the present work is at least able to contribute in some way to this end.

4. On the reasons for the very special value of mathematical symbolism, numerical as well as geometric, one may refer particularly to the explanations given in *The Reign of Quantity and the Signs of the Times*.

Index

www.ingramcontent.com/pod-product-compliance
Lightning Source LLC
Chambersburg PA
CBHW021932190326
41519CB00009B/993